Iran's Nuclear Program
Realities and Repercussions

IRAN'S NUCLEAR PROGRAM

REALITIES AND REPERCUSSIONS

THE EMIRATES CENTER FOR STRATEGIC
STUDIES AND RESEARCH

THE EMIRATES CENTER FOR STRATEGIC STUDIES AND RESEARCH

The Emirates Center for Strategic Studies and Research (ECSSR) is an independent research institution dedicated to the promotion of professional studies and educational excellence in the UAE, the Gulf and the Arab world. Since its establishment in Abu Dhabi in 1994, the ECSSR has served as a focal point for scholarship on political, economic and social matters. Indeed, the ECSSR is at the forefront of analysis and commentary on Arab affairs.

The Center seeks to provide a forum for the scholarly exchange of ideas by hosting conferences and symposia, organizing workshops, sponsoring a lecture series and publishing original and translated books and research papers. The ECSSR also has an active fellowship and grant program for the writing of scholarly books and for the translation into Arabic of work relevant to the Center's mission. Moreover, the ECSSR has a large library including rare and specialized holdings, and a state-of-the-art technology center, which has developed an award-winning website that is a unique and comprehensive source of information on the Gulf.

Through these and other activities, the ECSSR aspires to engage in mutually beneficial professional endeavors with comparable institutions worldwide, and to contribute to the general educational and academic development of the UAE.

The views expressed in this book do not necessarily reflect those of the ECSSR.

First published in 2006 by
The Emirates Center for Strategic Studies and Research
PO Box 4567, Abu Dhabi, United Arab Emirates

E-mail: pubdis@ecssr.ae
Website: http://www.ecssr.ae

Copyright© The Emirates Center for Strategic Studies and Research, 2006

Distributed by The Emirates Center for Strategic Studies and Research

All rights reserved. Except for brief quotations in a review, this book, or any part thereof, may not be reproduced in any form without permission in writing from the publisher.

ISBN 10: 9948-00-847-2 hardback edition
ISBN 13: 978-9948-00-847-7 hardback edition

ISBN 10: 9948-00-848-0 paperback edition
ISBN 13: 978-9948-00-848-4 paperback edition

CONTENTS

Abbreviations and Acronyms — ix

Foreword — xi
Jamal S. Al-Suwaidi

Introduction:
Iran's Nuclear Activities: Issues and Implications — 3

1. Iran's Nuclear Capability and Potential to Develop Atomic Weapons — 11
   John Simpson

2. Iran's Nuclear Program after the 2005 Elections — 37
   George Perkovich

3. Iran's Nuclear Program: Impact on the Security of the GCC — 63
   James Noyes

4. Israel and the Strategic Implications of an Iranian Nuclear Weapons Option — 93
   Geoffrey Aronson

5. Bombing Iran: Is it Avoidable? — 113
   Sverre Lodgaard

Contributors — 141

Notes — 145

Bibliography — 157

Index — 165

ABBREVIATIONS AND ACRONYMS

AIPAC	American Israel Public Affairs Committee
AWACS	Airborne Warning and Control System
CONUS	Continental United States
CTBT	Comprehensive Test Ban Treaty
DPRK	Democratic People's Republic of Korea
GCC	Gulf Cooperation Council
HCoC	Hague Code of Conduct against Ballistic Missile Proliferation
HEU	high-enriched uranium
IAEA	International Atomic Energy Agency
LEU	low-enriched uranium
mbpd	million barrels per day
MEK	Mujahedin-e-Khalq
MTCR	Missile Technology Control Regime
NFZ	Nuclear-Free Zone
NNWS	non-nuclear weapon state
NPT	Non-Proliferation Treaty
NSG	Nuclear Suppliers Group
PLO	Palestine Liberation Organization
SWU	separative work unit
UN	United Nations
UNMOVIC	UN Monitoring, Verification and Inspection Commission
UNSC	UN Security Council
WMD	weapons of mass destruction
WTO	World Trade Organization

Foreword

Following years of clandestine activity, the Islamic Republic of Iran's nuclear program has been referred to the United Nations Security Council, as a result of the continued diplomatic pressure and persuasion applied by the United States and the EU-3 (the United Kingdom, France and Germany).

In January 2006, when Iran began to ignore limitations on nuclear fuel cycle activities, the International Atomic Energy Agency (IAEA) and the international community adopted a tougher stance towards the Islamic Republic. Iran now appears to be at the beginning of the process of nuclear development and production, its short-term aim being to acquire options with regard to a nuclear capability. By initiating this process, Iran has defied the broad opinion of the international community and has insisted on continuing its program of nuclear research and materials enrichment despite both offers of assistance and the threat of international sanctions.

Iran has an historic aspiration to achieve the status of a great power in the region. If it were to develop a nuclear capability, this would provide the current Iranian regime with an important advantage in this regard. Therefore the world now faces the prospect of the Islamic Republic of Iran becoming a nuclear power and posing a potential threat to regional and global stability. Iran already constitutes a significant military force in the Gulf region, and it is therefore important to assess how far an Iranian nuclear weapons capability could be developed if the Islamic Republic were to choose this path.

The importance of finding a diplomatic solution to Iran's nuclear row becomes more apparent when considering the potential costs of military action. The issue may draw US forces further into the region, and there can be little doubt that any US or Israeli military strike targeting Iran's nuclear facilities would further complicate the overall security situation in the Gulf.

In view of the significance of Iran's nuclear program and its possible future repercussions on regional and global security, the Emirates Center for Strategic Studies and Research (ECSSR) hosted a symposium entitled "Iran's Nuclear Program: Realities and Repercussions," which took place on February 26, 2006, in Abu Dhabi, United Arab Emirates. At this symposium, visiting experts were invited to share their views on the Iranian nuclear program, its likely development and the extent to which the potential threat of nuclear weapons could destabilize the Arabian Gulf and the broader Middle East region.

This volume is a valuable collection of these expert views, covering current and future trends pertaining to Iran's nuclear program; Iran's current nuclear capability and its potential to develop atomic weapons; the changes and developments in the country's nuclear program since the 2005 Iranian presidential elections; and the possible impact which an Iranian nuclear program could have on GCC security.

An Israeli military strike on Iran remains a possibility and is a move which would entail severe regional political and strategic implications. Therefore, this book also assesses Israeli foreign policy in relation to Iran; examines the role of regional diplomacy in addressing the issue of the Iranian nuclear program; and explores the prospect of military action against Iran.

ECSSR would like to take this opportunity to thank the authors for their valuable contributions. A word of thanks is also extended to ECSSR editor Francis Field for coordinating the publication of this book.

Jamal S. Al-Suwaidi, Ph.D.
Director General
ECSSR

INTRODUCTION

Iran's Nuclear Activities: Issues and Implications

International suspicion surrounding Iran's nuclear activities was first aroused in the late 1980s and early 1990s.* However, it was only in 2002, when satellite photographs revealed undeclared nuclear facilities under construction, that the IAEA launched an intensive investigation into the Iranian nuclear program. One of the most alarming revelations of this investigation was that Iran had enriched uranium and separated plutonium in undeclared facilities in the absence of IAEA safeguards. Since these revelations, global concern has grown, largely due to events during the summer of 2004, when Iran breached the terms of a 2003 suspension by producing and assembling centrifuges, and in January 2006, when it crossed the clearest of international "red lines" by resuming work on uranium enrichment at its Natanz plant.

In June 2005, the first round of presidential elections in Iran saw the relatively unknown, firebrand politician Mahmoud Ahmadinejad gain the upper hand over the former president Hashemi Rafsanjani. This victory, coupled with the ongoing Iranian nuclear activities, deepened widespread concerns over the true nature of the Islamic Republic's indigenous nuclear energy program and the potential

---

* This introduction is based on the contents of the papers presented at the ECSSR symposium entitled "Iran's Nuclear Program: Realities and Repercussions."

effects of a nuclear Iran upon regional and international security. The main concern is that modern uranium enrichment technology has an inherent dual-use nature. With some modifications, a plant dedicated to producing fuel for nuclear power stations is capable of being used to produce materials that are indispensable in the production of a nuclear explosive device.

In his chapter entitled, "Iran's Nuclear Capability and Potential to Develop Atomic Weapons," Professor John Simpson explores in detail why the US and other states have come to regard the operation of an enrichment plant by Iran as a "red line" that it should not be allowed to cross. He points out that the optimum course for a modern proliferator might be to pursue a "smart" proliferation strategy of remaining within the NPT; acquire dual-use enrichment, reactor and reprocessing facilities; and stockpile sufficient low-enriched uranium or weapons-grade plutonium for conversion into 10–15 conservative fission devices. Once this point is reached, the 90-day notice of withdrawal from the Treaty could be offered, after which "breakout" could be implemented without breaching any legal agreement. Logically, the next technical steps after withdrawal from the NPT would be to convert the stockpiled low-enriched uranium into high-enriched metal and produce a few conservatively designed "emergency weapons" similar to those dropped on Japan in 1945.

Simpson argues that any assessment by Iranian technicians – or those analyzing the situation from outside Iran – of the time it would take to possess a fully weaponized force of significant size and technical credibility will probably begin with a target date of 2013–14 and move on from this. However, a force based on an "emergency capability" could be in existence from around 2009 onwards.

Since the collapse of Saddam Hussein's military forces in Iraq, Iran has achieved unquestioned conventional military superiority in the Gulf region. Iran's geographic advantages and large population of some seventy million dwarf the GCC states. In evaluating the effects of an Iranian nuclear program on the GCC, James Noyes stresses that the issue of Iran's nuclear activities is too dangerous for the GCC to dismiss. In his chapter entitled, "Iran's Nuclear Program: Impact on the Security of the GCC," Noyes points out that Iran's strident and defiant fervor conveys direct military threats towards the GCC as well as the West. This is not only because the United States is closely associated with nearly every aspect of the GCC's own military programs, but because Iran has demonstrated a host of new conventional weapons with potential uses against the GCC. Determining whether these weapons would perform as claimed is just as difficult as making a more fundamental assessment of the regime's motivations. Nonetheless, their presence cannot be ignored when the GCC calculates its own defensive capabilities and political vulnerabilities vis-à-vis Iran.

Direct or proxy Iranian assets are adjacent to Saudi Arabia's sensitive, largely Shiite Al Hasa region in the Eastern Province. Therefore, with Shiite minorities in many of the other GCC states – and Bahrain with a majority Shiite population – GCC spokesmen have usually muted their concerns about Iran's conventional weapons programs. However, concerns over the issue of the nuclear program have been less muted. This is partly due to the GCC's apprehension over the potentially catastrophic environmental implications of a nuclear accident at the Russian-built nuclear reactor at Bushehr, which is only 150 miles across the Arabian Gulf from Kuwait. With Iran's notoriously earthquake-prone geology, plus the reportedly Chernobyl-era technology of the Bushehr reactor, the possibility of an accident is alarming. An earthquake could cause an accident that would affect Gulf countries more than Iran. Such a catastrophe could kill 200,000 people and pollute the Gulf to the extent that the six water desalination plants on the Arab shore would be forced to shut down.

In terms of future strategies which might be considered when dealing with Iran's nuclear program, Dr. George Perkovich, in his chapter entitled "Iran's Nuclear Program after the 2005 Elections," points out that the first strategic imperative in dealing with the Iranian nuclear issue is to change the domestic dynamic in which Iranian leaders debate their nuclear policies.

In Iran, a range of actors and institutions are involved in nuclear policy-making, who wrestle with each other to reach a consensus. So far, the consensus has been that Iran should seek to acquire the capability to enrich uranium and (with less certainty) to separate plutonium. Iran is also developing ballistic missiles in ways that suggest an interest in being able to use them to deliver nuclear weapons. Therefore, to alter the domestic decision-making environment in Iran, the negative consequences of proceeding with a nuclear fuel-cycle program must be rendered much more severe, while the benefits of switching to international nuclear fuel services should be dramatically increased.

The second imperative is to create a broad international coalition to reinforce the certainty of both the consequences associated with a confrontationist policy and the increased benefits of a cooperative nuclear policy. Any compelling coalition must include Russia, China, India and Japan in addition to the United States and Europe. President Ahmadinejad and other autarchic Iranians have argued that Iran does not need to seek integration with the West and that the EU-3's leverage in nuclear negotiations can be circumvented by turning away from Europe and dealing with the East—in particular China and perhaps Russia and India. It is thought that these states will be more concerned with developing energy relations with Iran than promoting nuclear nonproliferation and human rights. Therefore, to alter Iran's aggressive nuclear policy, its champions must learn that they cannot "escape" to the East, and that the coalition to persuade Iran to accept international nuclear fuel services includes all of the world's major powers.

To further cement the coalition of countries urging Iran to adopt less threatening nuclear policies, the EU-3 and the United States should seriously consider the objective of creating a Middle East WMD-free zone. James Noyes highlights the fact that as the proliferation issue has moved to the forefront of international politics because of confrontation in the Arabian Gulf, new opportunities have arisen for the GCC to lobby for a nuclear-free Middle East. Although the problem of devising a verification system which Israel would find convincing appears insurmountable, it is not beyond the capacity of the world's scientific community.

Possible nuclear proliferation resulting from Iran's program is discussed on the sidelines of the military and environmental issues demanding GCC attention. Saudi Arabia, considered the most likely next nuclear developer, has categorically denied any such intentions. It would appear, however, at least over time that a nuclear-armed Shiite Persian Iran would constitute an unacceptable challenge to the Sunni Arab guardians of Mecca and Medina. With ample financial resources and a historically close military relationship with Pakistan, Saudi Arabia would presumably be able to arm itself with nuclear weapons short of lengthy and inevitably visible technological development.

Iran's protestations of peaceful intent must be weighed not only against IAEA inspections, but against the recent history of both India and Pakistan's successful covert nuclear weapons development. Estimates of when Iran might produce a bomb vary from ten or twenty years, to two years or within months if components were imported. The possibility of US military action on the other hand is conceivably imminent while an Israeli strike seems much less so. If the publicly available assessments are to be believed, Israel, which has had ample provocation from Iranian President Ahmadinejad, could not by itself easily reach Iran's numerous dispersed nuclear development installations. Distance, coupled with the number of sorties required would stretch Israel's assets unreasonably.

There is currently no regional power with the capacity to pose an existential threat to Israel, but Israeli policymakers consider any Iranian effort to gain a nuclear weapons capability as a move which establishes Iran as a potential existential threat—particularly in light of the Iranian regime's vitriolic statements against Israel. In his chapter entitled "Israel and the Strategic Implications of an Iranian Nuclear Weapons Option," Geoffrey Aronson explains that the Israeli decision to create a dedicated second strike capability presumes that Israel's nuclear monopoly will soon be ended and replaced with a bipolar or multipolar nuclear Middle East in which Israel is threatened by nuclear-armed ballistic missiles. In this environment, nuclear deterrence – and the creation of a second strike capability should deterrence fail – would become a favored Israeli option. The new environment created by the Iranian efforts may also hasten the day when Israel's nuclear capability is brought out of the shadows, where it has been kept not so very well hidden for more than a generation. This policy – which once served Israeli interests so effectively – has been overtaken by events. In order to protect the credibility of its nuclear arsenal in a nuclear environment, Israel may decide to adopt a policy of "open deterrence."

Concerns about Iran's efforts in the nuclear domain have led to a growing international campaign aimed at short-circuiting not only illicit Iranian weapons activities but also activities not prohibited under the Nuclear Non-Proliferation Treaty (NPT). Israel has supported these efforts – indeed it has championed them – but if the international community mobilizes against Iran, there is also a chance that the intrusive gaze of the international community will focus on Israel's own nuclear capabilities as part of a revived effort to establish a regional diplomatic framework for nuclear disarmament. There was a hint of this in the February 4, 2006 IAEA statement calling for a Nuclear-Free Zone (NFZ) in the region. Israel is not prepared to enter into a framework that would establish any sense of symmetry between its nuclear weapons and those of its regional antagonists. While Israeli–Iranian rapprochement cannot be dismissed as impossible, current affairs are more polarized than at almost any time since the Iranian revolution.

A US attack on Iran's nuclear facilities – unfortunately in public minds potentially confused with Israel's one shot destruction of Iraq's Osiraq facility in 1981 – would require a major exercise of war. Given the precise location, degree of hardening and depth underground of Iran's dispersed facilities, an attack would require many sorties and would have to be preceded by extensive repression of Iranian air defenses, including sophisticated Russian and North Korean surface to air missiles, anti-aircraft guns and fighter aircraft as well as command and control facilities. James Noyes quotes a former head of the US Defense Intelligence Agency's Middle East section, who predicts a US assault would require around a thousand sorties, involving air strikes and cruise missile operations, followed by multiple re-strikes. Such attacks would occur over a number of days and would have to include a concentrated campaign to preempt Iran's undoubted reprisal efforts which would be aimed at the United States' GCC allies as well as US targets in the Gulf. An Oxford Research Group study published in February 2006 concluded that a US attack aimed at setting back Iran's nuclear program by at least five years would also have to target Revolutionary Guard facilities close to Iraq and irregular naval forces that could disrupt Gulf oil or transit routes in retaliation.

A preemptive US attack on Iran would therefore present the GCC with a devastating scenario. Given no option but to sustain its basic defense relationship with the United States, the Arab Gulf states would be exposed to critical physical as well as political damage. Even if Iran's immediate reprisals against US military facilities – particularly in Bahrain, Qatar and the UAE – were minimal, an extended Iranian reprisal effort could be anticipated. This reflects the core problem: the extensive and lengthy US military action required for the operation would ultimately constitute a war between Iran and the United States. Such a war would draw US military forces even deeper into the region, fortifying extremist anti-western ideologues and jihadists. These consequences would not be confined to the immediate Gulf region but would extend throughout the Muslim world, which is already profoundly divided in its efforts to reconcile the pulls of modernity with traditionalist demands.

In his chapter entitled "Bombing Iran: Is it Avoidable?" Sverre Lodgaard writes that the advantages of halting the current escalation and achieving a political solution can be fully assessed only when compared to the potential costs of war. The costs cannot be predicted with any precision but they are potentially huge, not only for the warring parties but for the entire Middle East region. A war would also take its toll on energy prices and economic development worldwide. Compared to the case of Iraq – where realistic assessments of the long-term consequences of war were absent – much attention has been drawn to the political implications of the use of force against Iran.

The United States has always been ready to put pressure on Iran, and has been moving towards an escalation of the conflict, especially since August 2005, when Iran rejected the offer of the EU-3. It has been argued that far from being a foregone conclusion, the United States has so many other problems on its hands that, if anything, it would try to avoid a military outcome. Others point out, however, that what politicians do when they do not know what to do is buy time—and this is what bombing can achieve. Bombing would set the Iranian program back and buy time for efforts to change the regime. However, in a long-term perspective, bombing is likely to be counterproductive—for example, it was after the bombing of Osiraq that Saddam Hussein mobilized his resources behind a comprehensive nuclear weapons program.

Precision bombing could reduce civilian losses. Nevertheless, such losses will be large, for many targets are located in or near cities. In the aftermath, the nuclear inspectors will no longer have access to facilities and the nuclear program will be restarted. If Iran was not motivated to build bombs before, it will be determined to do so afterwards. Signs of renewed activity may trigger new bombing raids similar to those in the no-fly zones over Northern and Southern Iraq after the Gulf War of 1991. This is a recipe for continued high tension in the region and will block initiatives for confidence-building and arms control.

1

Iran's Nuclear Capability and Potential to Develop Atomic Weapons

John Simpson[*]

Many states are now of the opinion that Iran is seeking to acquire nuclear weapons. This belief is based largely on circumstantial data, as no hard evidence or "smoking gun" has yet emerged to provide convincing proof that Iran has an active nuclear weapons program. This circumstantial evidence relates to Iran's behavior and likely motivation. From a motivational perspective, Iran finds itself bordering two nuclear-armed states—Pakistan and Russia; since 1979, has been in a situation of political friction with a nuclear-armed United States whose military has a global reach; was the subject of major chemical attacks on its ill-protected troops and missile attacks on its unprotected cities by its neighbor Iraq in the 1980s; and – if its President's words are any guide – its current regime regards itself as being in acute religious and ideological conflict with a nuclear-armed Israel. Iran also regards itself as a state with deep historic Persian roots; was once a proud and great power and intrinsically remains so; and aspires to re-emerge as one again—in a regional if not a global context. Above all, many of its citizens see themselves as having a right to make the transition from being primarily a producer of a key raw material – oil – to a modern, advanced industrial state.

---

[*] The author is grateful to Jenny Nielsen for her research assistance in producing this paper.

Initially known only to national intelligence organizations, the circumstantial behavioral evidence takes the form of perceived attempts by Iran to import – openly or clandestinely – materials and technology which could be used to create the infrastructure for a nuclear weapons program. After 2002, this evidence entered the public domain, as a result of the International Atomic Energy Agency's (IAEA) desire to document and understand the full extent of Iran's nuclear activities. The IAEA failed to find any hard evidence of a nuclear weapons program, but did establish that Iran had acquired the technology to fabricate centrifuges suitable for the enrichment of uranium – and possibly samples of the centrifuges themselves – clandestinely through irregular non-state procurement channels. The Agency also ascertained that Iran was in the process of building both pilot and industrial-scale plants to operate these centrifuges, which were capable of producing – if required – uranium of suitable enrichment for the core of a nuclear explosive device.[1]

Opinions differ on whether Iran was legally required to declare these activities to the IAEA under the terms of the safeguards agreement in force at the time of the acquisitions. However, the clandestine nature of its procurement activities, their non-state character and Iran's unwillingness to offer a full account of its activities at the start of the investigation, have all served to reduce confidence in Iranian claims that its intentions are peaceful in nature and aimed at achieving long-term energy security through the acquisition of a complete nuclear fuel cycle. The latter aim has also contributed to this lowering of confidence in Iran's claims, as seeking a complete fuel cycle appears to make little economic sense at this stage in the country's nuclear energy development.

The widespread concerns over the true nature of Iran's indigenous nuclear energy program and the potential effects of a nuclear-armed Iran upon regional and international security have intersected with a more global issue: the future effectiveness of the nuclear non-proliferation regime based upon the 1968 Treaty on the Non-Proliferation of Nuclear Weapons (NPT). The concern is that uranium enrichment technology – especially of the modern centrifuge type – has an inherent dual-use nature. With some modifications, a plant dedicated to producing fuel for nuclear power stations is capable of being used to produce the materials that are indispensable in the production of a nuclear explosive device.

The dual-use nature of such plants was recognized at the inception of the NPT and in the design of its linked system of IAEA safeguards. The latter were intended to provide assurances to other states that enrichment plants within a state were not being used for weapons proliferation. However, the initial implementation of IAEA safeguards focused solely upon the diversion of materials from declared to other uses: only in the 1990s did they begin to directly address the possibility of clandestine activities. Yet a year after the "model" safeguards agreement between an NPT non-nuclear weapon state and the IAEA had been finalized in 1973, many of the states which possessed nuclear technology and materials had concluded that these verification arrangements would not prevent a determined proliferator acquiring the material to make nuclear weapons.

Those who already possessed the technology therefore sought to reinforce the NPT and the IAEA safeguards arrangements through a nuclear export control regime—now known as the Nuclear Suppliers Group (NSG). This operated nominally to determine which items were to be subject to export controls, but it also appears to have included a *de facto* ban on the transfer to states outside the NSG of those sensitive uranium enrichment, fuel reprocessing and plutonium separation technologies capable of being used in the development of weapons. As a consequence, the NSG and its activities were opposed by many developing states on the grounds that they hindered their economic development; were discriminatory; and were contrary to Article IV of the NPT, which legitimizes the peaceful use of nuclear technology. For two decades, however, these export control arrangements appeared to have provided an effective barrier to overt nuclear proliferation and the further dissemination of dual-use technologies.

The concerns which led to the creation of the NSG were, however, reaffirmed in the 1990s by the consequences of the theft by A.Q. Khan of centrifuge enrichment technology from the European uranium enrichment consortium Urenco. Initially this technology was used to provide Pakistan's national nuclear weapons program with fissile material. However, it appears that by the mid-1990s the clandestine, sub-state, multinational nuclear weapon supply network that Khan had created was also being used to provide at least two states – Libya and Iran – with a fissile material production capability on an apparently commercial basis. These developments led to frictions and questions over the nature of the contemporary NPT-based nuclear non-proliferation regime. Was it essentially a political construct to influence overt and public nuclear weapons-related behavior, but not clandestine nuclear weapon activities? Or was the aim to physically prevent such activities and deny states the ability to proliferate—in a non-discriminatory manner if possible, but if not, in a discriminatory one? The latter interpretation can be seen in the current US policy of seeking to prevent any further construction of enrichment plants on a global basis—possibly by force if necessary,[2] but preferably by making them unnecessary for civil power purposes by guaranteeing supplies of fuel to those who require them on a multinational basis.[3]

It is at this point that the global non-proliferation policies of the US and some other Western states intersect with their specific regional policies towards a possible Iranian nuclear weapons program. By denying Iran the right to develop and operate an enrichment plant, they are pursuing both a global and Iran-specific policy at the same time. If the linked multinational supply and enrichment control initiatives focused on Iran fail, it will both enhance threat perceptions in the Gulf and generate negative global consequences in terms of proliferation.

However, the political problem that exists is that Iran does not currently posses nuclear weapons, and thus does not pose an immediate nuclear threat in the Gulf—the criteria that would normally be regarded as justification for a military response. Instead, it is merely thought to aspire to develop a dual-use capability that *could*, under certain circumstances, provide it with the capability to make nuclear explosive devices and/or nuclear weapons systems. Arguably, this would place Iran in a similar position to that of Japan (and perhaps other advanced industrial states such as the Netherlands and Germany), but it could also give the country an ability to "breakout" – to develop nuclear-weapons possession at short notice – and/or a capacity to develop a clandestine dedicated nuclear weapons program in parallel with its declared and safeguarded civil one.

The remainder of this paper will therefore explore in greater detail why the United States and other states have come to regard the operation of an enrichment plant by Iran as a "red line" that it should not be allowed to cross. It will do this by examining what is actually involved in possessing a nuclear weapons "capability," as opposed to having the proven – or believed/claimed – ability to detonate a nuclear explosive device. It will examine the risks that Iran will face in going down this road and the strategies it may adopt to achieve these aims. Finally it will examine the nexus between these strategies and the concerns over Iranian nuclear-related actions that are at the root of current international concerns.

The Basics of Nuclear Weaponry

Nuclear weapons technology is now over 60 years old. Since its inception, knowledge of the nature and effects of atomic weapons – including the processes required to produce such weapons and credible systems to deliver them with certainty – has slowly disseminated, although many secrets remain. This dissemination particularly applies to general information regarding initial designs of fissionable material production plants and weapons developed by acknowledged nuclear-armed states. Yet although component technologies may have changed and short-cuts to designs have become available through transfers of technology – whether purposeful or inadvertent – the basic requirements for a nuclear-weapons program that existed in earlier periods remain relevant. Above all these include the need to bring together multidisciplinary teams to design, engineer, produce, maintain and operate such weapons.

Atomic weapons are not like cannon balls or bullets, which have no moving parts and are capable of being stored ready for use indefinitely. The first US weapons were described as having the complexity of a small aircraft, and it would be normal for such first generation weapons to be serviced – and in some cases remanufactured – at relatively short intervals as knowledge and engineering experience developed. The South African nuclear arsenal, for example, was remanufactured a number of times during its relatively short existence. Therefore, any discussion of Iran's potential for acquiring nuclear weapons must inevitably begin by making a few distinctions and definitions concerning the basic nature of such devices and weapons.

The term "nuclear" is usually taken to cover all applications of nuclear energy. These applications involve two possible processes: the fission and fusion of atoms—the building blocks of the universe. A nuclear explosive device or weapon based solely on fission is known as a *fission* or *atomic* weapon; one which generates a large proportion of its yield from fusion is known as a *thermonuclear* or *hydrogen* weapon. There can also be hybrid devices using both processes which are situated between these two polar extremes.

Nuclear Weapon Materials

Very few materials are capable of fission or fusion. Those atoms that are fissile (i.e., can be split) have a very heavy mass (i.e., weight), while those capable of fusion have a very light mass. Both heavy and light types of atom exist in different forms known as isotopes, distinguishable by their slightly different masses. The materials used to make fission devices are uranium-235 (U-235 or high-enriched uranium) and plutonium-239 (Pu-239). Those used to make fusion devices are lithium-6 deuteride and the hydrogen isotopes deuterium and tritium. Without access to fissile materials a state cannot make a fission or fusion device; without access to fusion materials it cannot make thermonuclear weapons or hybrid devices.

Uranium occurs naturally and is mined in many locations, including in Iran. U-235 is produced by separating the 0.7% of U-235 from the other isotopes found in natural uranium. To do this, uranium ore is first converted into gaseous uranium hexafluoride (UF_6—a very corrosive and dangerous material). It is then processed in a plant equipped with "cascades" of precision-engineered centrifuges spinning at very high speeds. These enable the material to be gradually enriched as it moves from one cascade to the next. The result is a pyramidal process, where the majority of the work has to be done in moving from the natural level of 0.7% enrichment to 5% enrichment, rather than from 5% to 93%—the preferred level of enrichment for nuclear explosive devices and submarine reactors. Finally, for use in an explosive device, the gaseous high-enriched uranium hexafluoride has to be converted into metallic uranium, whose only applications are military. It was Iran's efforts to acquire the necessary technologies to perform these operations that enhanced concerns about it having a clandestine nuclear weapons program.

Plutonium is produced when natural or slightly enriched uranium fissions in controlled conditions within a nuclear reactor. These conditions involve the presence of material to cool the reactor, and to slow down or moderate the subatomic neutron particles that are one of the products of the reaction. One of the best moderators is heavy water (ordinary water enriched in deuterium), although most power reactors use ordinary water (light water). Pu-239, the fissionable isotope, is produced by irradiating uranium in a reactor for a relatively short period of time, and then separating the plutonium from the uranium fuel chemically. Again, it is Iranian plans to acquire such technologies in a non-transparent manner that have generated international concern.

Nuclear Weapon Designs

The key characteristics of a nuclear weapon in comparison with all other types of weapon lie in its combination of power, instant effect and unique technology. Its power is usually described in kilotons, that is, in terms of its equivalence in conventional explosive power. This is because when nuclear weapons were first developed in the 1942–5 period, one way of explaining to politicians their implications was to compare their effects to that of the explosion of the *Mont Blanc*, a ship loaded with 4,000 tons of TNT in Halifax, Nova Scotia in 1917. Therefore, this 4 kiloton explosion, and photographs of the large area of blast damage it created, became the benchmark for contemporary descriptions of the anticipated power of an atomic bomb.

This single figure is also used to encompass the thermal and electromagnetic effects of such a device—even though such an aggregate figure is by its very nature imprecise. In addition, such explosions will also produce radiological fallout that may make territory inaccessible for periods of time unless those wishing to enter have suitable protective equipment. There may also be radiological and genetic effects on those individuals exposed to the direct radiation generated by such a device.

The majority of these effects take place almost instantaneously, rather than through the relatively lengthy processes characteristic of many biological and chemical weapons. Nuclear devices/weapons are thus qualitatively different in their effects from all types of conventional weapons and from other weapons of mass destruction (WMD). In practice, the contemporary yardstick for such devices is not the Halifax explosion but the blast and radiological/genetic consequences of the nuclear bombs dropped by the US on the Japanese cities of Hiroshima and Nagasaki in 1945. These "emergency" weapons, with a power of between 3–5 times that of the Halifax explosion, killed tens of thousands of people in an instant.

The unique technology that characterizes a nuclear weapon involves not only the two distinct processes, fission and fusion, but also the differing means of achieving them. The common element is the ability to compress nuclear material into a very small gaseous mass, which then supports a self-sustaining nuclear reaction for milliseconds until it blows apart. Two types of devices are feasible in the case of fission or atomic weapons: implosion and gun. In the former, high explosives are wrapped around a core of fissile material, usually spherical, to generate a converging implosion shock-wave; in the latter a small piece of U-235 is fired at a larger one placed at the end of a thick tube similar in nature to a gun barrel.

In implosion devices, the same explosive yield can either be obtained by using significant quantities of high explosives and a small amount of fissile material or less high explosives and more fissile material (i.e., increasing or decreasing the amount of compression). Thus, a weapon with a low yield but relatively light weight may often contain more fissile material (and therefore be less "efficient") than one of higher yield containing less material. This is important for assessing the number of nuclear devices a state can produce from a given amount of fissile material. In a hypothetical Iranian case, for example, more fissile material would be needed if it was to be used for a lightweight missile warhead than if it was to be used for a much heavier bomb or static demolition charge.

The operational advantage of a gun design is that it is simple in nature and can be used to create more compact and robust devices suitable for applications such as artillery shells, missile warheads or earth penetrators. The disadvantage of gun designs is that only U-235 can be used in them, and they are less efficient in converting fissionable material into yield than implosion devices. Thus, fewer gun weapons than implosion devices can be produced from a given quantity of this scarce material. However, the gun device dropped on Hiroshima in 1945 was not tested prior to use as it was regarded as more reliable than the implosion device dropped on Nagasaki, which was tested. Although South African devices used the gun technique, Pakistani devices use implosion, as did the designs transferred to Libya (and possibly Iran) by A.Q. Khan. Implosion designs are inherently more complex and less robust than gun designs, but are able to produce much higher yields. Test devices have been exploded with yields of up to 720 kilotons,[4] but for reasons related to the spherical nature of blast effects, most operational weapons would normally have a yield of 15–60 kilotons.

Thermonuclear devices, by comparison, have in theory no upper limit to their yield. Classic thermonuclear devices are double bombs, where the radiation yield from a first device serves to generate very high compressions in the second. No state is likely to be able to deploy these without explosive testing, and a proliferator would only graduate to thermonuclear devices after some years of experience in constructing increasingly smaller fission devices. Thus, unless Iran was to choose to break the current testing moratorium and repudiate its signature of the Comprehensive Test Ban Treaty (CTBT), this is not an option which is available to it.

Finally, having a nuclear weapons capability has a wide range of possible meanings and interpretations. There is a significant distinction between developing the art of fissile material production and weapon design and having the actual article. A nuclear device capable of being exploded underground may demonstrate that a state has mastered the initial stages of nuclear weapon design. However, for such a device to be used as a weapon within a military context, it has to be capable of being stored for years; be relatively reliable in use; be safe from accidental explosion in storage, transport and immediately before use; and have effective safety and security devices so that it cannot be stolen or misused by groups outside the control of the possessor state.

In addition, it has to be mated with a delivery system which has a credible possibility of reaching its assigned target. This means establishing the technical credibility (and safety within the controlling state) of the delivery system; the ability of its nuclear warhead to withstand the environmental and other effects of the delivery experience; and its ability to explode at the correct height and location. Of course if a nuclear device was to be used by a terrorist group, and either smuggled into a target country or assembled there, the environmental requirements would be much less demanding than in the case of a weapon intended for use by a national military force.

Practical Constraints on an Iranian Nuclear Program

Nuclear proliferation has never been simple. This is still the case, and Iran will not find the task easy if it chooses – or has already chosen – to move down this path. In the 1960s the international community opted to actively seek to limit the number of nuclear-armed states by constructing both legal and voluntary constraints to make the task of potential proliferators as difficult as possible. These barriers have evolved over time and expanded in their scope and reach. Therefore, any Iranian nuclear weapons program would either have to overcome or outmaneuver them.

Legal Constraints

Legal constraints start with membership of the NPT, which may impose serious restrictions on a state seeking to proliferate. All non-nuclear weapons states in the Middle East region – with the exception of Israel – are members of the NPT, and thus all their facilities containing nuclear material are open to IAEA safeguards inspections. This means that they could, in theory, produce HEU and Pu-239 under IAEA safeguards. However, from an economic perspective, Iran's lack of power reactors – other than the one under construction at Bushehr, for which fuel will be provided by the Russian Federation – makes it difficult to argue that such production would be solely for civil purposes. Instead, it is likely to be viewed as a form of "smart proliferation." This is a term used to describe a situation where a state remains in the NPT until it has sufficient fissile material to make 10–15 nuclear devices and then gives its mandatory 90-day notice of withdrawal from the Treaty.

The CTBT has been signed by many states, and ratified by some, but has not yet entered into force. Of the states in the Middle East, Syria and Iraq have not signed; Jordan, Kuwait, Oman, Qatar and the UAE have signed and ratified; and Bahrain, Egypt, Iran, Israel and Yemen have signed but not ratified. This makes nuclear testing a difficult decision to take for those states in the latter two categories, as it will almost certainly open them to legal arguments that their actions are illegitimate—based in part on the Vienna

Convention on the Law of Treaties. This could in turn result in the imposition of sanctions by the UN Security Council (UNSC). Also, there is a global moratorium on nuclear testing and the international reaction to any breach by a state such as Iran is likely to be extreme hostility and mandatory UNSC sanctions.

Export Controls

There is a well developed global system of guidelines for nuclear exports, designed to prevent external assistance reaching potential proliferators. One element of this involves listing those items which should only be exported by a state if IAEA safeguards are applied to them (the Zangger list). A second list (that of the NSG) contains those items which should be subject to export licensing (and thus potential denial) by the states involved. These lists indicate items their authors regard as necessary for a nuclear weapons program, the transfer of which should therefore be closely scrutinized.

In terms of delivery systems, the vehicle of choice for nuclear weapons is now a ballistic missile. Export controls on such missiles and their components are operated through the Missile Technology Control Regime (MTCR), to which the majority of missile technology holders subscribe, as well as the Hague Code of Conduct against Ballistic Missile Proliferation (HCoC). While indigenous development of missiles is possible, those pursuing this course have to overcome the difficulties involved in developing effective guidance systems; lightweight but strong missile structures; large rocket motors and effective and accurate re-entry systems. Without significant imports of technology, expertise and materials from other states, such development programs are likely to be lengthy, expensive, involve many tests to demonstrate the reliability of the systems (both to a state's own government and others), and be open to defensive measures by target states.

The Problems of Weaponization

Other lists exist which focus more specifically on the components a state would need to acquire, or the actions it would need to undertake to weaponize a nuclear device and make it operational by mating it with a specific delivery system. Both states and international organizations use such lists as warning devices to initiate more detailed monitoring of a state's nuclear activities and evaluations of its progress towards developing a capability to manufacture and deploy an operational nuclear weapon. These national and international lists focus on three main areas: weaponization, delivery and induction.

Weaponization is the process of moving from an initial design of a nuclear explosive device to a warhead which is safe to handle, can be stockpiled for long periods of time and can survive the journey from its storage area to a target. Weaponization activities involve:

- explosion theory (materials, experimental facilities and technology);
- neutron initiator experiments (materials and technology);
- fissile material theory (experiments and technology);
- tamper and tamper-movement theory (experiments and technology);
- gun and implosion assembly theory (experiments and technology);
- detonator and explosive experiments (materials and technology);
- warhead structure, carcase and casing design and engineering;
- fusing, arming, firing and safety system design and testing;
- vibration and environmental testing; and
- component engineering and manufacture.

The second area involves the manufacture of a credible delivery vehicle – including design, development, testing and production – or its acquisition by transfer. As many first generation ballistic missiles, such as Scud variants, are very inaccurate over anything other than short ranges, nuclear or other WMD warheads are the only ones which appear cost-effective for use with such a delivery system. Thus, efforts to procure such missiles with a range of over 150–200 miles are often regarded as indicators of a nuclear weapons program, as their inherent inaccuracy will be such that they only seem to be worth deploying as military weapons if they carry such a warhead.

The third area follows on from the second and is the induction of a nuclear armed delivery system into military service. Indicators of this would be the construction of bases and nuclear warhead storage facilities with their associated security systems, as well as personnel development in terms of safety, security, system training and dispersal exercises. In the case of aircraft-delivered weapons, training exercises would include practicing the special maneuvers a carrier aircraft would need to take if it were to escape the blast effects of its delivery of a nuclear bomb. Concern over Iran's nuclear intentions arises therefore not only from its nuclear activities, but also from the fact that it is seeking to develop – and by implication deploy – a medium-range missile system that could be used to deliver a nuclear warhead.

It is far from clear, however, what type of delivery system a nuclear-armed Iran would require. Much would depend on the type of targets it would want to attack or threaten to attack, and the defenses that could be deployed to defend potential targets against specific delivery systems. The simplest form of delivery system could involve attaching a gravity bomb to an aircraft (military or other), or the insertion of some type of "sneak weapon," dropped into a harbor or carried in a freight container. In these cases, technical issues such as the weight of the warhead, its centre of gravity, its shape and whether it has the degree of robustness and hardening which would enable it to survive a journey into near-space and back to earth would have less significance. However, if Iran sought to develop a device capable of being delivered to a target via a gravity bomb or missile, while offering a technically credible nuclear deterrent through this means against a state some distance from its borders, this would be a much more demanding and time consuming activity in comparison to a device which could be used for terrorist purposes, or delivered in a non-military manner.

To offer some idea of what might be involved in the process of weaponization, the first British bomb design, code-named *Blue Danube*, underwent over 200 drop tests on its bomb casing and ballistics; detonators and fusing systems; in-flight insertion and other safety systems; and its ancillary equipment before it could be declared fully fit for operational service. These tests took almost five years to complete, *after* the first nuclear device was exploded. Even then, the reliability of the weapon could only be expressed in terms of the percentage of units that would not explode as planned due to inherent weaknesses in the design. British designers also had to confront the possibility that it could be rendered inoperable by potential defensive measures, and therefore had to develop means of overcoming them. Ultimately, to have a capability, or even a nuclear explosive device, is a long way short of actually having a weapon capable of being delivered with credibility by a military formation.

Nuclear Proliferation and Iran

There are many variations of what constitutes a nuclear capability that can be used for political leverage or military deterrent purposes—e.g., China has been known to claim that Japan is already a nuclear-armed state, as it has the capability to produce both U-235 and Pu-239, and is presumed to have the technology to design and produce a bomb or missile warhead within a matter of months. India and Pakistan had nuclear devices for over a decade before testing and declaring their existence. Israel has never claimed to have nuclear weapons, but there is circumstantial evidence of this from the existence of its un-safeguarded, large "research" reactor at Dimona. Iran does not currently have a nuclear device or the material to make one—unless it has successfully hidden nuclear weapon activities from the international community. A key question, however, is whether the possession of a safeguarded fissile material production capability will give it a nuclear deterrent capability of the type possessed by Israel. If not – which seems likely – what strategies are open to it to obtain such a capability? This will probably have to involve moving to a situation where others would infer that it had nuclear weapons.

Judging by the experience of other states, at least three types of capability would enable a state to be accepted as having this type of *de facto* nuclear weapons deterrent:

- an opaque or recessed capability (e.g., Israel now or India prior to May 1998);
- a confirmed capability—by either claiming its existence or exploding a nuclear device (the DPRK now or India and Pakistan after May 1998); or
- a transparent capability—achieved by weaponizing a device and overtly discussing its deployment (India and Pakistan by 2000).

Taken together, the restrictions on transfers and the Treaty constraints make it very difficult for NPT non-nuclear weapons states such as Iran to aspire to anything other than an opaque capability. However, the existence of the A.Q. Khan network has demonstrated that shortcuts may now exist to achieving a deployed capability. Arguably, the extent of such an opaque capability is a fission warhead that cannot offer the assurance of having been fully tested—although extensive non-nuclear explosive testing would offer a high probability of an implosion design functioning effectively. Prudence suggests that such a design would probably be very conservative in its nature and thus expensive in its use of fissile materials, even if blueprints for a less conservative design had been acquired. As a UK expert on the subject once commented about the experience of building US designs in the UK, "blueprints are the problem, not the solution."

Without nuclear explosive testing, it is difficult to see how a state such as Iran could develop and produce either a hybrid or thermonuclear design. Also, without explosive testing, uncertainties would remain over whether its designs were capable of withstanding defensive systems designed to exploit its possible inherent weaknesses and render it inoperative.

The optimum course for a modern proliferator may therefore be to pursue a "smart" proliferation strategy of remaining within the NPT; acquiring dual-use enrichment, reactor and reprocessing facilities; and stockpiling sufficient low-enriched uranium or weapons-grade plutonium for conversion into 10–15 conservative fission devices. Once this point is reached, the 90-day notice of withdrawal from the Treaty could be offered, after which "breakout" could be implemented without breaching any legal agreement. In parallel it would be possible to work clandestinely on the design of the warhead and overtly on the acquisition of a suitable delivery system. However, discovery of information on both activities would heighten concerns about a state's long-term intentions, especially if it were to involve development of a long-range missile capability with limited accuracy. Indeed, fear of discovery might deter such extensive 'pre-breakout' activity, especially if the expected results were sanctions or military action.

The next logical technical steps after withdrawal from the NPT would be to convert the stockpiled low-enriched uranium into high-enriched metal, and produce a few conservatively designed "emergency weapons" similar to those dropped on Japan. Explosive testing could then be implemented in order to perfect an advanced design, perhaps using some thermonuclear materials. These would use reduced amounts of fissile material to generate the same yield and allow a significant stockpile to be achieved more rapidly. In this way the inevitable period of extreme political and military vulnerability for a proliferating state using this strategy might be shortened, although not avoided.

An alternative strategy for a proliferator is to operate a clandestine weapons program and then conduct a test, or withdraw from the NPT, without having stockpiled sufficient fissile material to make a number of weapons. Such actions do not seem to constitute a prudent strategy, as they would leave the state vulnerable for a number of years to preemptive political and military action, possibly authorized by the UN Security Council, without any obvious counterbalancing security benefits.

Iran's Nuclear Energy Program

Currently Iran is still in the early stages of the nuclear weapons development and production process (if that is what it is seeking to pursue). In all probability its short-term aim is similar to that of France through to the late 1950s, namely to create dual-use options that can form the basis for a weapons development program if the decision to implement one is taken some years in the future. Judgments on Iran's aims are thus being made largely on the basis of its perceived intentions, not its proven capabilities. However, the IAEA's investigations suggest that since the mid-1980s, Iran has concealed from the IAEA and the world a number of activities which could assist it to acquire weapons-grade fissile materials and to build an explosive device.[5] What cannot be denied, however, is that its clandestine activities have not generated confidence among other NPT parties in Iran's peaceful intentions.

IAEA investigations since 2003 have confirmed that Iran has made a number of attempts to acquire the means to make fissile material—how purposeful this activity was is unclear as it seems to have stopped and started several times, and may have been partly driven by offers of assistance from external sources.[6] The activities involved have included both seeking to acquire the technology for the centrifuge and laser enrichment of uranium, and the production of plutonium from a heavy-water cooled and moderated reactor. However, the latter is some distance away from completion. Its more immediate significance could be as a source of material that could be placed in the core of a nuclear device to initiate an explosion by generating a burst of neutrons. Although the Russian Federation is building a large power reactor for Iran at Bushehr, it is unclear how this could or would fit directly into any military program.

Iran has built and operated a plant to convert uranium ore, mined indigenously, into uranium hexafluoride (UF_6). Operations at this plant were suspended in 2004 but have now restarted. It appears that some UF_6 was also obtained from another country. Blueprints and components for first- and second-generation centrifuges based on designs stolen by A.Q. Khan and modified for use in the Pakistani nuclear weapons program were acquired by Iran via his procurement network, and in 2004 work was in progress to build a pilot centrifuge enrichment plant at Natanz, as well as a separate, much larger one.

By 2003, a number of estimates existed concerning the status of Iran's alleged program and possible dates for its acquisition of the fissile materials for a weapon and the weapon itself. In December 2001 the US National Intelligence Council reported that most US intelligence agencies assessed "that Tehran could have one [nuclear device/"emergency" weapon] by the end of the decade, although one agency judges it will take longer."[7] In February 2003 the US Defense Intelligence Agency suggested that "Tehran will have a nuclear weapon within the decade."[8] A UK-based author reported at that point that, "The British and French governments estimate that Iran could have a nuclear capability by 2007."[9] Finally, David Albright and Corey Hinderstein, in a detailed analysis of Iran's capabilities published in September 2003, argued that "by the end of 2005, this plant [the Natanz enrichment plant] could produce 15–20 kilograms of weapon-grade uranium, enough for a nuclear weapon" and therefore that, "In the worst case, Iran could have a nuclear weapon by the end of 2005."[10] They appeared to be working on the assumption that it would take about 18 months from plant commissioning to having enough HEU for a bomb, and that weaponization was occurring in parallel to enrichment. In short, the estimated timescales at that stage varied between 18 and 84 months and beyond for Iran to make a nuclear explosive device, depending on whether the estimates were based on worst case assumptions of a flawless program and existing clandestine activities or those that might be regarded as more "realistic," and allowed for the inevitable technical development problems and material shortages that such a program would encounter.

Estimates started to change following Iran's November 2004 suspension of its enrichment activities. Albright and Hinderstein, writing in January 2006, at the point when enrichment activities were restarting, believed that Iran possessed 700 centrifuges capable of being used in centrifuge cascades (for quality control reasons, many of those produced had been rejected), and that it had the capacity to manufacture 70–100 centrifuges per month at that point.[11] In a worst-case scenario, they concluded that by the end of 2006 Iran could have 1300–1600 usable centrifuges; by the end of 2007 Iran could have placed them into cascades, installed control equipment, built feed and withdrawal systems and tested a complete enrichment plant; would have sufficient HEU to make a device by the end of 2008; and could deploy it "in 2009." Thus their original worst-case estimate of 18 months to a device had extended by early 2006 to 36–48 months, as more detailed information had become available through IAEA reports and other sources on Iran's centrifuge program and the technology involved in it.

By March 2005 the US Defense Intelligence Agency had revised its estimate from its 2003 position of "within the decade" to "early in the next decade,"[12] which suggested a 60–72 month period to assemble a device. In October of that year, however, Henry Sokolski argued that Iran was still only "12–48 months from acquiring a nuclear bomb"[13] (i.e., 2006–9). More conservatively, Joseph Cirincione suggested that Iran is "five to ten years (60–120 months) away from the ability to enrich uranium for fuel or bombs" (i.e., 2011–2016) and asserted that this view was backed by a range of experts elsewhere.[14] Finally, at the end of March 2006, on the basis of briefings made to permanent members of the UNSC by the IAEA, Albright and Hinderstein offered a more detailed analysis of the worst case clandestine and "LEU conversation to HEU" Iranian breakout scenarios, which suggested that the earliest time to a nuclear weapon by either route was 2009.[15] Of perhaps greater significance, however, it suggested that if the LEU conversion route were to be adopted, Iran's breakout time for a single device could be 1–12 months, though it offered no figures for a stockpile of 10–15 devices.[16]

The reason these estimates differ so significantly probably relates to differing assessments of the difficulties the Iranians will face in enrichment and weaponization, and the assumptions concerning the context in which enrichment is to occur. Centrifuge enrichment is an inherently difficult engineering and control operation, involving building and operating a complex plant using dangerous feedstuffs, high speeds and close tolerances. It also involves the precision engineering of centrifuges using special materials and alloys which may be in short supply. This makes it almost inevitable that technical problems and material supply bottlenecks will slow down and delay the program, as has happened elsewhere. Thus the ability to make a small amount of low-enriched uranium in a pilot plant is only a first step towards reliably producing kilogram quantities of

HEU on an ongoing annual basis. The South African enrichment plant, for example, broke down after an initial period of operation and it was over a year before it could be restarted.

It is unclear at the moment from IAEA or other sources to what extent Iran has undertaken active work on weapon design. It does appear that work was undertaken in Iran on polonium,[17] which has few uses other than in initiators for an early generation fission weapon, but Iran would probably be unable to make this material for military purposes in a reactor while IAEA safeguards were in place. Iran also appears to have been offered equipment by outsiders that would enable hemispheric shells of HEU to be constructed, together with ideas on how to fit a nuclear warhead into a missile. Circumstantial evidence arising from the A.Q. Khan network's activities in Libya suggests that Iran could also have been offered blueprints based on China's fifth nuclear device. This was tested on a Chinese missile and modified by Pakistani weapon designers before being copied and sold abroad by Khan.[18] Yet even if this is true, it is unlikely that such a design would have been slavishly copied by Iranian weapons designers in the absence of extensive experimentation to prove it was not part of a disinformation exercise.

However, it is not unreasonable to assume that by early 2009 a weapon design could be proved by non-nuclear explosive testing and that conversion, casting and machining of HEU hemispheres and other weapon components could have started. In these circumstances, a device might be capable of being assembled within 6 months. However, the result would be an "emergency deliverable weapon," rather than a fully weaponized and tested missile warhead. Arguably, the latter could take another five years to develop, especially if the initial missile design had to be altered to accommodate the physical characteristics of the nuclear warhead and the warhead design itself had to be developed further to reduce its weight and so increase the reach of its delivery system.

One final variable is that after commissioning an enrichment plant, Iran would have to decide whether it was going to leave the NPT and move directly to make HEU: or remain in the NPT, produce low-enriched uranium suitable for reactor fuel in the plant, and stockpile it under IAEA safeguards for rapid further enrichment to HEU at a later date. The former approach could lead to a device being completed sooner than the latter. However, completion of a single device is not in itself the acquisition of a nuclear deterrent, other than possibly against non-nuclear armed neighboring states. Rather it represents the opposite: the point of greatest insecurity once its existence is known or declared. Production of further devices would only increase slowly as more centrifuges and cascades came into operation—assuming no military strikes or material bottlenecks.

Moreover, a number of years would probably elapse before a missile-based weaponized nuclear force could be deployed. Yet even that could be countered by moves from Iran's neighbors to adopt the type of arrangements made in Europe from the mid-1950s onwards (i.e., allowing nuclear-weapon states which Iran cannot easily target to base their nuclear weapons on a neighbor's territory), or the deployment of defensive measures designed to exploit the known vulnerabilities of early generation nuclear weapon systems. Thus, even at the end of the lengthy period leading to the deployment of a missile-based deterrent with some ability to survive a disarming strike, it would still be likely that any prestige or regional security advantages gained at great expense from Iran's production and deployment of nuclear weapons would be outweighed by the enhanced threats to its national security that would result.

Conclusion

What, therefore, can one conclude from this situation? First, Iran does not currently have a nuclear weapon and its capability to produce one is not yet well developed. Indeed, despite its announcement of March 11, 2006 that it has a new status as a "nuclear" power given its claimed enrichment of uranium, this offers little in the way of hard evidence to persuade other states that it currently possesses an opaque nuclear weapons capability, let alone an operational one. The situation is similar to someone who has a bullet claiming that they have a handgun or rifle. Secondly, if Iran goes ahead (or is allowed to go ahead) with its large uranium enrichment plant, but remains in the NPT, it will have to decide whether to leave the Treaty and move directly to high enrichment, or to enrich to a level below 5% and store the material produced in order to give it a reasonably rapid "breakout" capability.

Thirdly, if Iran leaves the NPT and goes straight ahead with high enrichment, it might take only 18 months in ideal conditions from that point to have enough U-235 to produce its first device, and perhaps another 24–36 months to deploy a credible stockpile mated to a gravity-bomb delivery system, with a missile force taking significantly longer. However, conditions will be far from ideal as Iran will undoubtedly be subject to a regime of escalating sanctions and military counter-proliferation threats, if not actions, as well as more rigorous import restrictions. Thus development, production and deployment periods are likely to be long.

Fourthly, if Iran chooses to stay in the NPT and stockpile low-enriched uranium, it could have a credible capacity to "break out" into nuclear-weapon status and produce several weapons within 12 months of any eventual decision to withdraw. It is the fear of this scenario of "smart proliferation" that appears to be motivating the US and other states, including Israel, to regard the start of enrichment as "the point of no return" or "the crossing of the red line," as the time needed for converting a stock of LEU into the 20kg of HEU required for a device/bomb could be as little as 1–3 months. Once Iran has perfected the "art" of indigenous enrichment, the option of moving down this road will always be available to it, as will its ability to transfer its technology to others. Denial of this art is thus only possible at this point and if large-scale enrichment proceeds, a breakout to nuclear-armed status may be too rapid to prevent by diplomatic action alone.[19]

Finally, any assessment by Iranian technicians – or those analyzing the situation from outside Iran – of the time it would take to possess a fully weaponized force of significant size and technical credibility will probably begin with a target date of 2013–14 and move on from this. A force based on an "emergency capability" could be in existence from around 2009 onwards. Yet in either case, the nuclear force Iran deploys is likely to be subject to "escalation dominance" from existing nuclear-armed states, who could station nuclear weapons in neighboring territories or at sea with no credible military nuclear counter-threat to themselves. However, whether Iran could deter or neutralize such actions by asymmetric resource denial and economic threats is another matter.

There could also be internal doubts among Iran's nuclear engineers regarding the technical vulnerability to defensive countermeasures of any nuclear-armed force the country deployed. Thus, by any normal calculation of costs and benefits, it appears that Iran will lose security from moving forward with a nuclear weapons program, rather than gain it, for any attempt at breakout will mean having to survive a period of acute vulnerability to military action, perhaps lasting 3–5 years, before it has any hope of generating a credible regional deterrent—and even then this could be easily offset by nuclear guarantees from states outside the region. The opportunity costs to Iran of diverting resources into a missile/nuclear weapons project that would probably offer low reliability and doubtful technical credibility in reaching distant targets would also be significant. However, in the final analysis, this may not be the type of "Cold War" deterrence thinking and rational assessment of politico-military needs that is driving, and will continue to drive Iranian policy.

2

Iran's Nuclear Program after the 2005 Elections

George Perkovich

Act I of the Iranian nuclear drama began with the international community's discovery in 2002 that Iran had been shirking its commitments to the International Atomic Energy Agency (IAEA). The Islamic Republic of Iran's main antagonist – the United States – seized on the damning evidence and attempted to convince the rest of the world that Iran's violations proved that it was too dangerous to be allowed to produce materials that can be used in the production of nuclear weapons.

The drama over Iran's quest for nuclear technology intensified and appeared headed toward a climax with the entrance of President Ahmadinejad in August, 2005. Unable to provide answers to the IAEA that would allay suspicions that Iran's nuclear activities have not been entirely peaceful, and seeking to compel their neighbors and the world to respect Iran as the major power of Southwest Asia, the clique around Ahmadinejad chose to flout limitations on nuclear fuel-cycle activities in an overt display of bravado. This clique acts in a way that suggests that the Iranian people neither need nor want to be integrated into the international community—particularly the community of values and institutions associated with Europe. "Iran for the Iranians, and maybe the Chinese and Malaysians," appears to be the logic. Such coarse rhetoric and actions on the part of this clique have attracted the attention of many other countries. When the IAEA discovered yet more evidence that Iran was pursuing applications of nuclear technology that were not exclusively peaceful, a majority of the international community decided that Iran must be chastened and compelled to act in a more trustworthy fashion, so they referred Iran to the UN Security Council.

This paper will begin by summarizing the events from 2003 to August 2004, and proceed to explore the changes since the 2005 elections, analyze the underlying forces at play and suggest policies that the United States, the EU, Russia, India, and Iran's neighbors should consider in the future.

Iran's Nuclear Ambitions

Iran's nuclear activities first aroused suspicion in the late 1980s and early 1990s. However, suspicion became alarmed certitude in 2002 when satellite photographs of undeclared nuclear facilities under construction in Iran prompted the IAEA to launch an intensive investigation into the Iranian nuclear program. The subsequent investigation prompted a series of further revelations that since the 1980s Iran had consistently failed to comply with its obligations as a non-nuclear weapons state. One of the most alarming revelations was that Iran had enriched uranium and separated plutonium in undeclared facilities in the absence of IAEA safeguards. Moreover, because these and other activities were secret and contrary to its obligations, Iran had resorted to lies and deception in its reports to the IAEA. The pattern of Iranian activity and secrecy confirmed many observers' suspicion that Iran was seeking the capacity to enrich uranium (and possibly to produce and separate plutonium) in order to have – at the very least – the option to manufacture nuclear weapons. While the evidence collected thus far does not prove with 100 percent certainty that Iran seeks to acquire nuclear weapons, it does provide an overwhelming justification for doubting that Iran's activities are exclusively for peaceful purposes.

The IAEA was compelled by the facts to declare on September 24, 2005, that Iran's failures and breaches constituted non-compliance with its safeguards agreements. According to the IAEA statute (Article XII.C), when the Agency determines that a state has failed to comply with its obligations, the Agency's Board of Governors "shall report the non-compliance to all members and to the Security Council and General Assembly of the United Nations."

However, from the time of the 2002 revelations and the subsequent discovery of major Iranian violations, Iran made clear that it was eager to avoid having its case reported to the UN Security Council. In an informal crisis management process, it was agreed that three leading states of the European Union – France, Germany and the United Kingdom – would negotiate with Iran to find a way to satisfy the IAEA and the international community's interests in preventing nuclear proliferation. At the same time, they would also seek to serve Iran's interests, both in avoiding referral to the UN Security Council and in continuing its efforts to benefit from the peaceful application of atomic energy.

On October 21, 2003, Iran and the so-called "EU-3" reached an agreement whereby Iran would voluntarily suspend all enrichment-related and reprocessing activities. In return, the EU-3 – and by association the IAEA – would suspend the process of reporting Iran to the UN Security Council, as is normally required by the IAEA statute in such situations. The agreement stated that, "once international concerns ... are fully resolved Iran could expect easier access to modern technology and supplies in a range of areas," and that the EU-3 would "co-operate with Iran to promote security and stability in the region including the establishment of a zone free from weapons of mass destruction in the Middle East."

Of course, the term "suspension" connotes temporariness. The negotiations between Iran and the EU-3 were meant to lead to a permanent resolution of the problems associated with Iran's nuclear program. These problems were not fabricated by outside powers, but rather stemmed from Iran's extensive, documented violations of its nuclear obligations. Iran and the EU-3 recognized that the broader interests involving Iran included integration into international trade and the aforementioned goal of strengthening regional security—including addressing the conditions necessary to establish a WMD-free zone in the Middle East.

However, two enormous conflicts lay beneath the surface of this negotiating agenda. First, the EU-3, the United States and many others felt that Iran's alarming nuclear record, support of terrorist operations and organizations, and non-recognition of Israel, all meant that the world would not be secure if Iran acquired the capability to produce nuclear weapons fuel. Thus, their goal was to resolve the crisis in a way which would require that Iran met its nuclear energy needs without producing fissile material on Iranian soil. Iran, conversely, felt that its national security interests required the capacity to produce nuclear fuel on Iranian territory. "National security interests" should be understood broadly here—indigenous nuclear fuel production would protect Iran against nuclear fuel supply embargo, military threats and diplomatic blackmail, as adversaries would have to assume that Iran could make nuclear weapons. Iranian leaders also believe that fuel-cycle capabilities will bolster Iran's status as a major global player and the dominant player in the Arabian Gulf. Thus, Iranian and EU-3 aims were mutually exclusive.

Second, the EU-3 and many other states felt no urgency in concluding negotiations as long as Iran continued to suspend its nuclear activities. Iran, however, hoped to use the threat of ending the suspension as a means to press the international community to give in to its demands. While the EU-3 could have talked forever if

Iran would maintain the suspension, Iran's central objective was precisely the opposite—to end the suspension as soon as it was ready to proceed with developing its fuel-cycle capabilities.

It is important to note here that Iranian officials do not argue that Iran's security situation justifies the acquisition of a nuclear weapons capability, as some observers sympathetic to Iran claim. Rather, Iranian officials insist that their nuclear intentions are entirely peaceful and that Iran cannot be denied the "right" to nuclear technology, including specifically the capability to enrich uranium and produce plutonium. Of course, were Iranian officials to discuss why a nuclear weapons capability could be advantageous, it would be seen as an admission of a violation of its treaty commitments and would invite precisely the consequences that Iran is trying to avoid. Moreover, Iranian officials may genuinely believe that the country should not build nuclear weapons, even if it has mastered the fuel-cycle; they may believe that the very capacity to produce fissile materials is sufficient to deter adversaries and command the respect of neighbors and citizens, and therefore that talk of possessing nuclear weapons is superfluous—and indeed dangerous.

While Iranian officials (and most commentators) remain silent about the security risks and benefits of acquiring nuclear weapons, they overstate the "rights" that countries have to produce fuel that could be used in nuclear weapons. Clearly, countries have a right to benefit from atomic energy, yet this right – enshrined in the Nuclear Non-Proliferation Treaty (NPT) and in the basic conception of an international system of states with sovereign equality – does not specify rights to particular technologies. The NPT does not specify any particular technology to which compliant states are entitled or not entitled (except nuclear weapons, which are not defined). Such specifics are therefore subject to negotiation and evolving standards and practices. However, the treaty is very clear both in its purpose and its specific terms (Article II) that if a country seeks any non-peaceful application of nuclear technology, material or know-how it would negate its right to nuclear technology if that country was a non-nuclear weapons state under the NPT—as is the case with Iran. Given Iran's documented non-compliance with its nuclear obligations and insufficient explanations for facts that indicate military involvement in its nuclear program, there is no objective basis for accepting that Iran would only apply uranium enrichment and plutonium separation capabilities for peaceful purposes. This adds to the conclusion that the international community has a clear basis for insisting that a state in Iran's circumstances must negotiate precisely what type of nuclear technology it should possess. That is, the right to benefit from nuclear technology should be met in ways that protect international peace and security, which may include reliance on international fuel services in lieu of dual-use indigenous production.

Of course, much of this is disputed by the government of Iran. The fundamental differences between the interests of the EU-3 and Iran, and the disadvantages Iran felt under the suspension, made it inevitable that Tehran would soon seek to escape the framework it had accepted in 2003. Iranian negotiators and politicians regrouped and skillfully portrayed the crisis as an attempt by the imperial West to deny a developing Muslim country the rights to modern technology. They adeptly shifted the focus away from the undeniable facts that Iran had broken its commitments and still had not provided answers to outstanding questions over signs that its nuclear program was not entirely peaceful.

In the summer of 2004, Iran breached the terms of the 2003 suspension by producing parts for, and then assembling centrifuges. At first the EU-3 and other IAEA member states wrung their hands, while Iran said that these activities were not barred under the suspension as Iran understood it, and in any case, the EU-3 had not closed the Iranian nuclear dossier at the June 2004 IAEA meeting, so Iran was acting in protest. Gradually the EU-3 and the IAEA convinced Tehran that the Iranian dossier would be reported to the UN Security Council if Iran did not resume a complete suspension of activities related to uranium enrichment and plutonium reprocessing. In November 2004, days before the meeting of the IAEA board of governors, Iran once again voluntarily suspended its fuel-cycle activities.

The Nuclear Issue Following the 2005 Elections

The first round of presidential elections in Iran in June 2005 surprised much of the world—including perhaps most Iranians. The relatively unknown, austere, firebrand politician, Mahmoud Ahmadinejad gained the upper hand over arguably the most well-known politician in Iran—the former president Ali Akbar Hashemi Rafsanjani. With chameleon-like use of palette, Rafsanjani portrayed himself as a moderate, modernist keen to integrate Iran into the international economy. Ahmadinejad expressed defiance of the outside world and a determination to build Iranian strength through greater social justice and moralism. He was backed by the Revolutionary Guard and other autarchic, morally conservative organizations. Against the notoriously wealthy and politically promiscuous cleric Rafsanjani, Ahmadinejad was the candidate of change, of protest—he was not a cleric, nor was he rich or corrupt. He seemed genuinely committed to greater social justice, equality and conservative values. The Iranian people clearly wanted change – as they had when they elected the reformist, Mohammed Khatami in 1997 – and Ahmadinejad easily defeated Rafsanjani.

Sensing the rise of militant nationalists and autarchic elements associated with the Revolutionary Guard, Iran's chief nuclear negotiator, Dr. Hasan Rowhani, gave a remarkable four-hour interview to *Keyhan*, a leading conservative newspaper.[1] Rowhani clearly meant to defend his own reputation, but also to educate new uninformed actors and their backers about the complex nature of the nuclear issue and Iran's strategy. It is difficult to imagine that

Rowhani would have revealed so much if permission had not been granted by Supreme Leader Khamenei. Rowhani explained that Iran's nuclear policy was set by the consensus of key leaders and institutions, including the Supreme Leader, in a "Council of Heads." This Council consists of Ali Khamenei (Supreme Leader), Hashemi Rafsanjani (President, Expediency Council), the defense minister (then Ali Shamkhani), the President (then Khatami, now Ahmadinejad), and the National Security Council Secretary (then Rowhani, now Ali Larijani). Rowhani explained that the Council made all major decisions—including on the suspension of fuel cycle activities, and on which activities should be covered by such a suspension. The Council of Heads also established the so-called "red line," that Iran must retain enrichment capability. The implicit message Rowhani gave in the interview was that newly-empowered firebrands should not discard the negotiating strategy lightly.

Most importantly, Rowhani explained that the concessions which Iran *appeared* to make when it agreed to suspend particular nuclear activities actually cost the country nothing. This was because the activities which were suspended were those that Iran's nuclear technologists were not ready to pursue, and whenever they had overcome difficulties or were prepared to take new steps, they did so. As Rowhani argued, when Iran started negotiations with the EU-3 in September 2003, "there was no such thing as the Esfahan project" to produce UF_4 and UF_6 for use in centrifuges. "But as of today," Rowhani said in July 2005, "we have prepared and tested the Esfahan facility on an industrial level and produced a few tons of UF_6. On the surface, it may seem that it has been a year and nine months since we accepted the suspension. But the fact of the matter is that we have fixed many of the flaws in our work during this period."[2] Rowhani went on to say that work at Esfahan and Arak had not been slowed by the suspension. "When a certain activity was suspended," Rowhani boasted, "during that period we would concentrate all of our effort and energy on other activities ... The day when Natanz was suspended, we put all of our effort into Esfahan. Now that Esfahan is in suspension, we are fixing other existing flaws."[3]

It was therefore not surprising, given Rowhani's revelations, that in August 2005, Iranian leaders decided to unilaterally break the suspension and begin operations at the uranium conversion plant in Esfahan. The technologists had experienced major difficulties in their past attempts to produce UF_4 and UF_6 of sufficient purity to be introduced into centrifuges, and the suspension merely coincided with the time they required to improve their techniques. They were ready to try again in August, so decision-makers broke the suspension. This technical push was wrapped inside a diplomatic–political offensive to demonstrate to the EU-3, the IAEA and others that Iran would not negotiate away its self-proclaimed right to the fuel cycle.

Even if Ahmadinejad had not been elected, Iran would have resumed uranium conversion for three reasons:

- First, the collective leadership was determined to manifest Iran's nuclear rights and to advance the nuclear establishment's capabilities.
- Second, the EU-3, the IAEA and the key members of the UN Security Council appeared to be unwilling to impose major penalties on Iran, so the suspension could be broken at minimum cost to the country; if a strong reaction appeared imminent, the work at Esfahan could be stopped after having successfully assessed the latest techniques.
- Third, the EU-3 and (perhaps most importantly) the United States had not offered sufficient positive inducements to give Iranian decision-makers reason to conclude that their overall political, economic, and security interests would be better served by considering limits on indigenous nuclear fuel-cycle development.

The Ahmadinejad Variable

President Ahmadinejad is not the key decision-maker in the collective process of determining Iranian nuclear policy, but has been able to shout his way on to center-stage, which could have consequences that are difficult to predict. In September, Ahmadinejad went to New York for the 2005 World Summit and surprised UN Secretary General Kofi Annan and other leaders by delivering a vitriolic, confrontational speech. Most observers, noting that Iran had alarmed the world by breaking the nuclear suspension, had expected that the new president would attempt to dissipate alarm by sounding reassuring tones. In October, Ahmadinejad made a speech reminding his audience of what Imam Khomeini had said: "the occupying regime [Israel] must be wiped off the map ... this was a very wise statement. We cannot compromise on the issue of Palestine ... I have no doubt that the new wave that has started in Palestine ... will eliminate this disgraceful stain from the Islamic world." In December, Iran insulted Russia by at first ignoring and later slighting Moscow's offer to involve Iranians in enriching uranium jointly in Russia.

Ahmadinejad's words and attitude are a new independent variable in the experiment of Iran's nuclear interaction with the world. If his aggressive style increases his popularity at home, it will make it harder for other Iranian actors to pursue more temperate policies. If Ahmadinejad does not gain much popularity at home, but becomes celebrated elsewhere in the Muslim world, it would also complicate Iranian foreign policy, as popular Sunni Arab desires for a more confrontational posture, especially toward Israel, could conflict with Iranian national interests in avoiding international isolation. If the Iranian people and/or the most powerful countries in the world allied to penalize Ahmadinejad's aggressive stance, advocates of a more temperate policy within Iran could gain political strength.

Through December 2005 it appeared that Ahmadinejad's approach was leading to positive results for Iran. His logic was clear: "The Europeans are like barking dogs," he was purportedly overheard as saying during his visit to New York, "if you kick them they will run away." Whether or not the president actually said this, Iran's actions from August to January amounted to kicking the dogs. At first the strategy appeared strikingly successful—as Berlin, London, Moscow, Paris, Washington and other capitals refrained from biting back. Internationalists in Tehran, including accomplished diplomats who had been summarily fired by Ahmadinejad, were chastised for having been weak. In effect, the message was: "See ... you accommodated these dogs, but if we are tough and resolute we will get what we want. There is nothing they can do to us." The aggressive, world-be-damned conservatives were triumphant, while the internationalists (who are hardly liberal) were on the run along with the Europeans.

Then, on January 10, 2006, Iran crossed the clearest of international "red lines" by ending the suspension of work on uranium enrichment at the Natanz plant. For the EU-3 to ignore this would be tantamount to running away with their tails between their legs. Russia, already insulted, growled in anger and IAEA Director General Mohamed ElBaradei showed teeth that he had not bared in three years of intensive management of the Iran dossier. Not only had Iran challenged the world by resuming centrifuge-related work, Tehran had done little to answer the outstanding questions regarding its past non-compliance and was still not giving IAEA inspectors the immediate access they sought to facilities, original documents and researchers.

While technical reasons and diplomatic frustration probably impelled the collective leadership to resume enrichment-related work, the world associated the move with the aggressive nature of President Ahmadinejad. The new burst of nuclear activity, paired with the hostile rhetoric and manner of the president, genuinely alarmed the rest of the world. Ahmadinejad and the crossing of the enrichment red line finally did what the Bush administration had been struggling unsuccessfully to do: it made the world fear Iran's nuclear ambitions more than it fears the United States' designs for stopping Iran.

Something had to be done and the only viable possibility within the rule-based international system was for the IAEA to report the Iranian case to the UN Security Council. By statute, arguably this should have been done in 2003. But now that Iran has ended its suspension of fuel-cycle activities, and shown disdain for the rules and concerns of the international community, the matter belongs in the Council.

Strategies for the Future

In the remainder of this paper I will suggest strategic and tactical guidelines that I believe would best enable the international community to persuade Iran to modify its nuclear plans in a way which guarantees that its nuclear activities will be exclusively for peaceful purposes. This discussion reflects a view of Iranian interests, predilections and politics—the filter through which any proposed agreement must pass. It is important, however, to emphasize the fact that Iranian interests, predilections and politics change frequently and are extremely difficult to predict. This is compounded by my own distance from the subject.

The first strategic imperative is to change the domestic dynamic in which Iranian leaders debate their nuclear policies. As the Rowhani interview states – and experience validates – a range of actors and institutions are involved in nuclear policy-making, who wrestle with each other to reach a consensus. Thus far the consensus has been that Iran should seek to acquire the capability to enrich uranium and (with less certainty) to separate plutonium. Iran is also developing ballistic missiles in ways that suggest an interest in being able to use them to deliver nuclear weapons. The consensus seems to dictate that Iran should pursue these capabilities without violating international treaties. Iranian decision-makers have sought to explain past violations of the country's nuclear obligations as being compelled by external threats and have indicated that they will not be repeated. Iranian leaders therefore want to be cooperative enough to close the prosecutorial case against them and escape international humiliation and sanction without abandoning the acquisition of capabilities that would enable the country to quickly produce nuclear weapons. Publicly available information does not conclusively validate conclusions that Iranian leaders have decided to actually build and/or deploy nuclear weapons if and when they have the capability to do so.

The pivotal issue, or independent variable, is whether Iranian leaders place a higher priority on acquiring the capability to enrich uranium (or separate plutonium) than on any other objective whose attainment the international community can facilitate or deny—at this point, the answer appears to be "yes." The international community has failed to convey an adequate deterrent to persuade Iran to seek reliable international nuclear fuel services, rather than enriching uranium or separating plutonium; nor has the international community convinced the Iranian people that the magnitude and reliability of the economic, political and security benefits Iran would gain from altering its nuclear fuel-cycle plans are so great that the existing policy should be changed. The threats being posed are too implausible or weak, while the benefits being offered are too limited and uncertain to persuade this particular constellation of Iranian decision-makers to change course. Therefore, the balance of costs and benefits must be altered. In all likelihood, to have this new balance result in new Iranian policies, the constellation of Iranian decision makers must be changed.

The ascent of Mahmoud Ahmadinejad as President of Iran provides a major opportunity for outside actors to consider the plausibility and scale of actions against Iran if enrichment-related work continues and key questions about its past violations and future intentions remain. Ahmadinejad has raised alarm among the international community and at the same time intensified competition within the Iranian elite. If this alarm can be translated into a skillfully marshaled plan for international action to impose penalties on Iran if it does not change course, the pressure could enable competing actors to form a new consensus on nuclear policy that would be less aggressive. For this to happen, however, the international community must match its Ahmadinejad-stimulated will to penalize Iran with a plan to display greater and more concrete benefits that rival actors in Tehran could "win" for the Iranian people through a change in nuclear policy. It is precisely Ahmadinejad's aggression that makes US and other politicians reluctant (indeed, politically frightened) to offer rewards, and instead makes them emphasize penalties. Such an unbalanced approach is guaranteed to fail. It will not create incentives that Iranian political competitors can seize upon to change the elite consensus on refusing to trade technologically-unproved indigenous fuel production for reliable international nuclear fuel services.

The first strategic imperative is to alter the domestic decision-making environment in Iran by making the consequences of proceeding with a nuclear fuel-cycle program much more severe, while dramatically increasing the benefits of switching to international nuclear fuel services. The second imperative is to create a broad international coalition to reinforce the certainty of both the consequences associated with a confrontationist policy and the increased benefits of a cooperative nuclear policy.

Any compelling coalition must include Russia, China, India and Japan in addition to the United States and Europe. Autarchic

Iranians, including President Ahmadinejad, have argued that Iran need not seek integration with the West. The perception is that the leverage of the EU-3 in nuclear negotiations could be circumvented by turning away from Europe and dealing with the East—in particular China and perhaps Russia and India. It is thought that these states will be more concerned with developing energy relations with Iran than promoting nuclear nonproliferation and human rights. To alter Iran's aggressive nuclear policy, its champions must learn that they cannot "escape" to the East, and that the international coalition to persuade Iran to accept international nuclear fuel services includes all of the world's major powers.

In addition to unifying the major global powers, the most important way to influence Iranian decision-makers is for Arab states to make it abundantly clear that if Iran develops the capability to produce nuclear weapons they will need to ensure that their interests are not overwhelmed by a hegemonic Iran. In the face of a nuclear-armed Iran, the most likely means of restoring a regional balance would be to seek (privately) a tighter military coupling with the United States and the deployment of additional US military firepower in the region—including ballistic missile defenses of all types. To convey Arab concerns and how those concerns might lead to actions that Iran would rather avoid, a regional security forum should be convened. The situation in Iraq, concerns over Iran's nuclear ambitions and turbulence in the wider region all call for the creation of a regional forum where Iran, Iraq, the Gulf Cooperation Council countries and adjoining states can consider alternate scenarios for their future relations.

To further cement the coalition of countries urging Iran to adopt less threatening nuclear policies, the EU-3 and the United States should seriously consider the objective of creating a Middle East WMD-free zone. This objective is implicitly included within the agreed mandate of the EU-3/Iran negotiations and has been explicitly endorsed in enough major international forums to warrant attention. No one should expect significant material steps toward such a zone to be taken soon, but the major powers' blatant inattention to this goal undermines international – and particularly Arab – support for an active coalition to influence Iran.[4] The ascent of Ahmadinejad and his radical backers provides a major opportunity; if other Iranian actors, including the Supreme Leader, see that an aggressive nuclear stance by Iran unites the major powers of the world in a coalition to isolate, press and contain it, the need for an alternative policy will become more obvious. Ahmadinejad and the institutions that back him can then be blamed for having endangered Iran's standing in the world and for forcing more responsible decision-makers to pursue international cooperation. Again, such a shift is only conceivable if it is accompanied by more well-defined international benefits than have been offered thus far.

A vital benefit which should be among those offered is a highly reliable mechanism for assuring Iran and other states that they will receive cost-beneficial nuclear fuel services from international suppliers if they abstain from local enrichment or reprocessing.[5] Ideally, an international mechanism would include the permanent take-back of spent-fuel by suppliers so that the country importing the fuel does not have to undertake the great expense, environmental risk and political controversy of designating a site for nuclear waste disposal facilities. A country like Iran would not be 100 percent confident that international fuel supplies would never be interrupted, but confidence in the reliability of supply must be made greater than has been allowed by the United States thus far. Washington understandably – but mistakenly – refuses to separate nuclear cooperation from other issues. Thus, the United States is currently unwilling to assure Iran (or others) that it would not exercise its power, directly or indirectly, to block nuclear fuel exports to Iran – even if Iran upheld all negotiated nonproliferation conditions – if Iran at the same time were engaging in terrorism or other hostile activities that would warrant sanctions. This conditional linkage to other issues is one reason why the benefits being offered to Iran for giving up indigenous uranium enrichment are ineffectual in changing the course of Iranian domestic decision-making.

The UN Security Council could augment the attractiveness of international fuel supplies by calling on international suppliers to provide uninterrupted nuclear fuel services to Iran as long as the country refrains from constructing and operating uranium enrichment and plutonium separation capabilities. The Council should invite Iran and a commission of industry and other experts under the aegis of the IAEA to explore modalities for the most reliable, cost-effective provision of international fuel services to countries that forego indigenous fuel production. The United States would be unlikely to veto Security Council language to bolster unlinked nuclear cooperation with an Iran that eschews national fuel production.

Any strategy to alter the cost/benefit calculations that shape the debate among competing Iranian decision-makers must involve the prospect of losses to the nation's economic and military security and the regime's political security if enrichment continues. Alternatively, Iranians must see greater prospects in terms of gains in economic, military and regime security if they choose to rely on international cooperation.

The latter category – regime security – is the most difficult for Washington to contemplate. The Bush administration's democracy-promotion agenda is widely supported, which makes it nearly impossible politically for Washington to unequivocally reassure the current government in Iran that the United States will respect its sovereignty. Such a reassurance would be seen as "selling out" the citizens of Iran who want to reform their political system and make it wholly democratic. President Ahmadinejad's recent rhetoric makes it still more difficult for US officials to advocate initiatives to reassure the Iranian government of its security.

A coalition of major global powers would provide an important basis for a process of confidence-building and security assurance and would be vital to the United States. If the members of such a coalition conveyed assurances that the sovereignty of the constitutional government of Iran will be respected as long as it does not pose a threat to international peace and security, this would put extraordinary pressure on Washington not to object. Over time, as progress is made in the nuclear sphere, the area of terrorism and in regional security forums, the United States could more explicitly affirm its respect for the sovereignty and territorial integrity of Iran. Washington should not be expected to cease its calls for greater political freedom and respect for human rights in Iran, just as Iran should not be expected to withhold criticism of myriad US policies. Rather, what would be established is a *modus vivendi* on the most fundamental issues of national and international security.

Sanctions on Iran

This chapter began by discussing a general framework and the need for a broad coalition, and then proceeded to discuss the benefits that must be made clear to Iran. However, no such discussion would be complete without exploring the options for sanctions. It is important to clarify that no party is eager to pursue sanctions and that no attempt will be made to impose them without clear provocation by Iran—in the form of actions that heighten the sense of nuclear threat as long as Iran has not completely resolved all outstanding issues with the IAEA. The following are some tactical thoughts about sanctions that reflect a sense of the political filter through which Iran sees and hears the world:

First, US and other officials weaken their position by talking publicly about sanctions or suggesting that an attempt to impose them is imminent. For sanctions on Iran to be effective, they would have to be applied by all important investors and traders who deal with Iran. For that to happen, sanctions must be authorized by the UN Security Council under Chapter VII of the UN Charter, which makes specified sanctions mandatory for all states. For the Security Council to pass mandatory sanctions, Russia and China must be willing to refrain from using their veto. Russian and Chinese resistance to putting the Iranian issue before the Security Council will be calmed if they are provided with high-level assurances that Chapter VII sanctions will not be sought as long as Iran is acting to rectify its non-compliance with IAEA safeguard requirements and is not producing fissile materials outside the arrangements agreed by the UN Security Council (once the Council has taken the matter under its auspices). In other words, Russia and China must be confident that the United States is not trigger-happy.

Second, it would make little sense to threaten or to impose significant economic sanctions on Iran without having prepared the people of the sanctioning countries for the potential consequences. (Political sanctions such as denying Iranian athletic teams the opportunity to compete internationally; banning travel of relevant Iranian officials and their families; denying landing rights to Iranian aircraft, etc., are less problematic.) Thus far, this has not been the case.

Third, governments should now spend time determining which specific forms of sanctions would have the most direct impact on the most threatening Iranian actors, while best sparing the Iranian public. Models should also specify the costs that various sanctions alternatives would impose on the sanctioning countries. For example, Iran is heavily dependent on machine tools imported from Europe. Therefore the following questions should be raised: which tools are most important to institutions associated with the Iranian military or nuclear sector? Which firms from which European countries export these goods, and what is their value? How could adjustments be made or burdens spread so that the costs of sanctions will be most palatable? What is the capacity to focus sanctions on organizations associated with the Iranian Revolutionary Guards, and on entities or activities on which the Guards can profit by controlling smuggling to work around sanctions?

As a general proposition, the two most effective categories of economic sanctions would be on foreign investment and exports into Iran, as the country needs both badly. Barring investments to Iran does not deprive investors of in-hand gains, but rather would move them to opportunities elsewhere. Barring exports to Iran is more problematic, but again, if well-targeted, such sanctions can be sustainable.

Fourth, the possibility of an oil embargo must be discounted. People know that most of Iran's income comes from oil exports and assume naturally that the most potent sanction would be to embargo that oil. At that point the plot twists, because if the world embargoes Iran's oil, the price of the oil the world buys will increase. This will not be tolerated by the people and parliaments of the United States, Europe, India and other countries. Therefore, in reality, this sanction will not be applied. Then the conventional wisdom becomes: if we can't use the biggest sanction lever we have, then there's no way to change Iran's nuclear course with weaker levers—therefore it is inevitable that Iran will get the bomb.

This seems to be the logical chain that some commentators are following, but I believe it is flawed. We don't know how Iranian politics will be affected by the many steps that should be taken to change Tehran's cost-benefit calculus. The international community would be better off communicating that it has many such steps in mind and being modeled, and that it will be unified in taking them if Iran continues to refuse to satisfy IAEA doubts about its past activities and reassure the world of its present intentions by eschewing the acquisition of bomb-making capabilities. The international community should indicate that the steps it has in mind do not include an oil embargo, but rather a number of more targeted measures which are not in its best interest to discuss publicly.

Military Action Against Iran

Beyond sanctions, of course, the ultimate proliferation prevention action would be to destroy a country's nuclear infrastructure militarily or, even more problematically, to eliminate a government and try – or hope – to replace it with one that would not continue intolerable nuclear activities. No one doubts the US military's unrivalled ability to precisely destroy targets identified for it. However, the well-known problems associated with a military strategy to prevent Iran from acquiring nuclear weapons stem not from the capabilities of the American armed forces, but rather from the intelligence that determines their targeting and the consequences of their well-executed actions. Leaving the enormous intelligence problem aside, US policy makers should ensure that the potential consequences of military action are fully modeled. After an action or move is described, the United States should ask, "what happens next?"; "what might they do next?"; "what would we do then?"; and "after we do that, what would they do?" and so on.

As we are learning in Iraq, it is important to try to anticipate many moves and counter-moves in a sequence that would follow military action. Conducting this exercise in anticipation of a military attack on Iran quickly leads to the conclusion that one can have little confidence in how the "game" ends. This is regrettable: the world would be a much safer place if Iran's decision-makers concluded that, at the end of the day, they could be physically prevented from acquiring nuclear weapons at a cost that was readily bearable by the international community.

For now, the most that can be achieved is the establishment of a coalition of the United States, Europe, Russia, China, India, Japan and Iran's neighbors; a coalition that would be in a position to create a cost-benefit table that leads reasonable Iranians to the conclusion that the best deal is to master nuclear technology in a way that gives the rest of the world no cause to think Iran intends to build nuclear weapons.

In the event that Iranians read such a table differently, or decide that nuclear weapons have a much greater value than anything else, this strategy will have to be abandoned. In its place, the United States and other concerned states will then be forced to focus all their energy and might on ensuring that Iran does not use a nuclear weapons capability to threaten or blackmail others.

When will we know it is time to switch to a contain-and-deter strategy? When Iran has proceeded successfully to enrich uranium, without having resolved the outstanding questions concerning its non-compliance with IAEA obligations, the United States and other responsible actors must prudentially move toward a deterrent strategy. It is in the interests of all concerned to delay Iran's enrichment of uranium and speed its resolution of outstanding IAEA questions, as this combination of outcomes would give all actors more time to avoid the inherent risks of confrontation. My own view is that the international community and Iran will be more likely to find a mutual accommodation when new governments (with fresh mandates, less baggage and more energy) are elected in Paris, Washington and London, and when Iran's current executives have demonstrably failed to deliver the job creation and reform that the Iranian population requires.

Iran's current government is determined to be recognized as the dominant actor in the Arabian Gulf and broader Middle East. Rather than seek this status by mobilizing its exceptionally talented civil society and potential for fulsome democratic politics, the revolutionary government seems to prefer more threatening tactics. It will be necessary to rid Iranian actors of any illusion that they could use a nuclear weapons capability as a shield behind which to conduct terrorist operations or other low-intensity conflict, believing that fear of nuclear escalation would keep the United States, the Israelis or others from acting forcefully. The message must be, "You may have acquired nuclear weapons, and we don't like it, but we're prepared to live with it and leave you alone. However, if you act violently in anyway outside of your borders, or utilize nuclear threats to blackmail others, we have the resources and the will to dominate the escalation ladder."

Historically, the most difficult challenge with states that have newly acquired nuclear weapons is to deter them from seeking to use this power as a shield behind which to seek political-strategic gain and/or to conduct low-intensity aggression. Thus far, no state with nuclear weapons has been willing to risk countervailing nuclear strikes in order to conduct major aggression. This, however, is merely suggestive and does not mean that such forbearance and rationality are inevitable. This is only a partial consolation. History tells us that the real risk is that a nuclear-armed state will feel emboldened to attempt to project its power in more insidious ways.

The most recent example has been Pakistan. Since it acquired a basic nuclear weapons capability in 1987, Pakistan has exploited local opportunities in Kashmir (beginning in 1989) to foment drastically increased violence. After the 1998 Indo-Pak nuclear weapons tests, Pakistan infiltrated into the Kargil area of Kashmir to take territory held by India in the spring of 1999. The Pakistani military was clearly emboldened by its nuclear deterrent cover to markedly increase coercion against India.

Iran would seem even more prone to this temptation. Its most militant and powerful "security" agency is the Revolutionary Guard, the leadership of which is believed to control the Iranian nuclear program and remain animated by revolutionary ardor and contempt for international norms. Iran also lacks a well-structured, linear decision-making apparatus, unlike, say, China when its still-revolutionary government acquired nuclear weapons. More than any current possessor of nuclear weapons, the post-revolution government in Iran often turns to brinksmanship, sometimes of a seemingly irrational nature, in its negotiations. Understandably, all of this invites caution when estimating deterrents which will prevent a nuclear-armed Iran from increasing its coercion and subversion of adversaries.

It is reasonable to conclude that the most important imperatives in a strategy to deter and contain a nuclear-weapons-capable Iran will be intelligence and international cohesion. Iranian practitioners of violence would have to know that the United States and neighboring-states can detect aggressive Iranian action, or moves to utilize nuclear weapons, and that Iran will not "get away with" coercive diplomacy or actual violence to threaten neighbors or broader regional interests. The United States (and possibly others) would want to bolster deployments of military and other assets surrounding Iran to improve its capacity to respond immediately to any act of Iranian or Iranian-sponsored aggression. These assets would probably include theater ballistic missile defenses. For this threat to be convincing, the United States would need the cooperation of neighboring states to allow US operations, and would need to be internally resolved to act decisively against any violence conducted by Iran or its agents outside Iranian territory.

Conclusion

For the world to be spared the dangers and insecurities that would follow if Iran were believed to have the capability to produce nuclear weapons, its leading actors must present a united front to make it clear to Iran that its future will be much more positive if it were to avoid the costs and environmental dangers of an indigenous fuel cycle and instead rely on internationally-supplied fuel services. Iran must be made aware of the benefits of cooperation and the grave costs of confrontation, and it must see this from all angles—not only from the West.

Chief negotiator Hasan Rowhani defined the contest very clearly in his *Keyhan* interview. He said that when Iran's secret nuclear activities were first exposed, the world was united against Iran. "Almost all of Iran's economic activities were locked ... a decline in business activities pervaded the entire market and even ordinary trades were affected, because ... Iran's case was going to the Security Council ... [T]he option of America's military attack was not very unlikely."[6] Had Iran's case actually been reported to the Security Council, the dossier of violations would have been thousands of pages thick with no clarifications or corrections from Iran to ameliorate the reaction. However, in July, 2005, after two years of diplomacy (and the troubled US occupation of Iraq), Rowhani claimed, "The political consensus that had formed against Iran at the outset has completely broken down." The lack of unified international pressure on Iran has therefore encouraged decision-makers to break the suspension at Esfahan and then in January 2006, at Natanz.

Now the question – and the challenge – is whether the aggressive actions and rhetoric reflected by President Ahmadinejad and the Iranian nuclear establishment will unite the international community against Iran acquiring a nuclear weapons capability. If it does, then Iran's collective decision-making is likely to point back toward international cooperation. However, if the major countries of the world break ranks in the face of aggressive Iranian thrusts, the confrontationists represented by Ahmadinejad will vanquish their internal competition and the world will come to represent a much more dangerous stage.

3

Iran's Nuclear Program: Impact on the Security of the GCC

James Noyes

The Gulf region's recent military history is dominated by disastrous miscalculations in planning, many of which relate to the misinterpretation of the relative strengths of nationalism versus sectarian and ethnic forces within state boundaries. For example, Saddam Hussein expected the Arab population of Iran's Khuzestan province to support his attack into Iran that began the eight-year Iran–Iraq War in 1980. Instead, the Iranian Arabs fought as Persians. Similarly, Ayatollah Khomeini apparently expected the Shia of Southern Iraq to welcome his onslaught into the Faw Peninsula, where instead they defended Iraq, causing disastrous Iranian losses.

Unfortunately, one can count among these follies those of the US architects of regime change in Iraq, who became convinced by their expatriate advisors that once Saddam Hussein was gone, Iraqi nationalism would unite the country and ultimately eradicate the multiple sectarian and ethnic divisions within its population.

While representative of very different motivations and objectives, these military campaigns shared a common denominator. In each case, ideologically-driven assumptions dominated analysis to the point of intoxication, overriding the sober weighing of historical and strategic factors. The mounting stresses which characterize today's GCC security environment are dramatized by Iran's nuclear program.[1] Ultimately, however, how these stresses are resolved will depend less on the introduction of new weapons programs and more on how the sectarianism inflamed by war is managed.

With or without nuclear weapons, Iran's unstable leadership and the potentially chaotic situation in Iraq have created new threats to GCC security; and the role of the United States as the GCC's security guarantor of last resort has become more complicated due to Iraq's deterioration. With the entire industrialized world – including Russia, China and India – engaged in the issue of Iran's nuclear program, the GCC countries are in the eye of a political storm. From this perspective they strive for a balanced view and voice in the proceedings.

The most pervasive and powerful subsurface element throughout the spectrum of GCC threats is the struggle within Islam itself between so-called "modernists" and extreme traditionalists. The various forms of terrorism endangering the region are largely byproducts of this struggle and the sectarian tensions it has exacerbated.

The aggressive regional role of the United States is itself, of course, a reaction to the attack on the World Trade Center in September 2001 and to other attacks on its interests including the Beirut Marine barracks truck bombing in 1983 and the 1996 bombing of the Khobar Towers US barracks in Saudi Arabia. Initially, these events tended to be identified by the GCC states primarily as anti-US acts related to Israeli–Palestinian issues, from which they were immune. However, once local terrorist activity struck against Saudi Arabia and other parts of the GCC, the viewpoint changed.

Ranking Security Threats

When asked to rank Saudi Arabia's security threats in January 2006, Nawaf Obaid, Managing Director of the Saudi National Security Assessment Project in Riyadh, replied, "First and foremost is obviously terrorism and Al Qaeda. Unfortunately this has a domestic and a foreign component. Number two is what will happen in Iraq. Third is the Iranian factor; the Iranian nuclear program is less a priority than the existing environment of government in Iran."[2] Obaid cited President Ahmadinejad's recent remarks to King Abdullah in Mecca, "that he considers Saudi Arabia as Iran's alter ego in the region: Iran represents the Shiite world and Saudi Arabia the Sunni world. This is very dangerous talk, positioning Iran as regional leader. This is the logic of what they do in Iraq and the nuclear sphere."[3]

Obaid's assessment highlights the basic GCC security position. Since the collapse of Saddam Hussein's military forces, Iran has achieved unquestioned conventional military superiority in the Arabian Gulf region. Iran's geographic advantages and large population of some seventy million dwarf that of the GCC states. Given this existing advantage, one must question what gains – in regional terms – nuclear weapons would bring Iran beyond its conventional superiority. If Iran chooses to employ military intimidation to further its political or economic goals in the Gulf, how much psychological power would the addition of nuclear weapons add to its campaign? This question is moot. A nuclear Iran will engender a degree of awe in the Gulf, but also suspicion and resistance. Primarily, Iran's nuclear quest points initially towards deterring Western and particularly US interference in its domination of the Gulf; and complicating if not completely deterring a US military attack designed to facilitate regime change. In the long-term, Iran's nuclear aspirations have been made inseparable from national honor by the clerical regime.

The international furor raised by Iran's nuclear program arises not only as a result of the potential threat to GCC security but because of Iran's support, largely through Syria, of Hamas, Islamic Jihad, Hezbollah and other organizations operating in the Levant. The resulting confrontation with the West creates such potential dangers for global energy supplies that Russia, China and India have joined the US and Europe in pressuring Iran to change course. Whether diplomacy may be effective in deescalating the confrontation remains an open question. Iran's provocation has been so strident and deliberate that it seems reasonable to assume that the hard-line President has deliberately fueled the dispute, partly to deter rapprochement with the United States—a goal that is sought by many of Iran's reformist groups.

Beyond this, Iran would have difficulty in extricating itself from confrontation with the West – even if reformists replaced hard-line factions – for two important reasons: first, Iran would need to admit extensive falsehoods to the International Atomic Energy Agency (IAEA); namely, hiding its nuclear program from inspectors for eighteen years. Second, were it not for Iran's post revolutionary behavior in general and the inflammatory rhetoric of its current president, Iran would have a reasonable case on a regional basis for possessing nuclear weapons. This case has become embedded in the ethos of important segments of the population. To now retreat even from its aggressive promotion of non-nuclear weapons programs would be a strenuous political feat for a regime with a paranoia recently fed by India and Pakistan's development of nuclear weapons. Iran, like others in the region, lives with the risk that Islamist extremists might seize power from Musharraf's regime and gain control of nuclear weapons. For Tehran, the difficulty of a "retreat" from its claimed peaceful nuclear energy ambition would be further complicated by the precipitous nature of the climb down from the rigid rhetorical confines of the current confrontation.

Iran's Claim

By virtue of Iran's national validity as an ancient civilization, the size of its population, and its strategic geographic location, future leaders will be unlikely to accept a second tier regional power status. Even during the Shah's pre-revolutionary era, nuclear energy was an element in his vision of an economically powerful and militarily dominant Iran with naval power over the entire Indian Ocean. Prominent military specialists assert the "high probability that he started and maintained a covert nuclear weapons program."[4] The most advanced US and European land, air and sea weapons platforms were being acquired or sought and had India and Pakistan's nuclear weapons burst onto the scene during that period, a passive response by the Shah would have been unimaginable.

From the beginning, however, Iran has insisted that its nuclear endeavors are for purely peaceful purposes. According to Hussein Mosavian, deputy head of Iran's delegation to the IAEA, Iran's current rate of growth will require the domestic utilization of the country's entire oil production within a few years. This would virtually eliminate Iran's vital oil export income.[5] However, the "peaceful use only" claim, which is virtually impossible to verify, is weakened by several factors. Given its enormous natural gas resources, Iran's burgeoning energy requirements would seem more efficiently satisfied by building gas plants rather than nuclear reactors. Moreover, Iran's elaborate deceptions over many years to conceal critical aspects of its nuclear program from the IAEA would be inexplicable.

On an immediately pragmatic basis, Iran's attraction to nuclear weapons – or at least to building the expertise and infrastructure permitting their development at reasonably short notice – undoubtedly derives in part from a radically altered post-revolution threat environment. The former US alliance has been replaced by bitter animosity, tangibly demonstrated by US support for Iraq during the eight-year war. This mutual animosity is now repeatedly expressed by threats of regime change on the one hand and on the other by the destruction of Israel and implied retaliation against the West through oil exports and terrorist methods. Iran therefore stands alone without reliable allies, except for an isolated and militarily weak Syria, itself under multiple threats. In addition, Russia and China can apparently only be relied on to block the approach of the more punitive US and European UN Security Council measures.

In weighing Iran's rationale for having at least a foothold on the ladder to nuclear weapons capability, the GCC must acknowledge the magnitude of Iran's defeat by Iraq in the eight-year war. Iran's far greater population and strategic depth were insufficient to offset the technological help and benefits Iraq received from the United States. Critical elements of Iran's military infrastructure lacked spare parts and modernization; and Iran had only a limited ability to match the devastating barrage of Iraqi missiles fired at its cities. Based on the events of recent history and a realistic assessment of its ability to circumvent sanctions or pay for a full modernization of its military in a suitable timeframe, Iran could be expected to seek shortcuts to a deterrent capability. Moreover, Iran's residual fears of a resurgent Iraq could only have intensified during the years of persistent claims by the West's intelligence agencies regarding Saddam Hussein's development – and ultimately, possession – of nuclear weapons. For Iran, having been on the receiving end of Baghdad's chemical assaults during the war, Western intelligence confirmation in this regard was unnecessary.

Missile development with existing or readily acquired technology and a limited need for skilled manpower has provided a shortcut that is an adequate façade to bolster domestic political aims. Tehran has ballyhooed this missile development in proportion to the mounting fever of confrontation with the United States. Operation Desert Storm in Iraq in 1991 and the US dash across the desert to

Baghdad in 2003 dramatized the gap between Iran's mostly outmoded equipment and the high level of technology employed by US and British forces. This Iranian disadvantage would be particularly telling in the realm of air power, where aging Western-supplied aircraft and the low training level of pilots and support personnel would not match the US, British and Saudi forces they might confront. The latter powers, of course, operate with state-of-the-art high-technology electronic warfare systems and targeting capabilities that can be integrated with AWACS airborne control, all with refueling capability.[6]

Missiles to Deter a Superpower?

These and other factors add up to a plausible case – from Iran's standpoint – for acquiring a missile-based deterrent capability. However, coupling this with many years of covert nuclear development generates deep concern among the GCC states. At the heart of this concern is the judgment and wisdom of Iran's leaders, who have placed their relatively weak state – from an international perspective – directly in the cross hairs of the world superpower's military sights. Every US president from the Second World War to the present has stated as policy bedrock that the United States cannot allow the oil resources of the Gulf to fall into the hands of a hostile power. Iran has just observed two major US military actions against Iraq that were partly motivated in support of this policy, yet it has gone out of its way to flaunt its hostility to the United States and to goad Washington with outrageous military threats against Israel and the United States itself. The fact that the United States has not achieved *political* success in either Afghanistan or Iraq does not detract from its proven capacity for the rapid destruction of a target state's military infrastructure.

Iran's game is too dangerous for the GCC to simply dismiss by making allowances for President Ahmadinejad's rashness as an unrepresentative voice within a diversified and moderate Tehran power structure. Nor can the US diplomatic miscalculations in dealing with Iran during past years justify the fervor with which Tehran has steadily layered confrontation not only with Washington but with most of the world's industrialized powers. Iran's strident and defiant fervor, by implication if not intention, conveys direct military threats towards the GCC as well as the West. This is not only because the United States is closely associated with nearly all aspects of the GCC's own military programs, but because Iran has demonstrated a host of new conventional weapons with potential uses against the GCC. Whether these weapons would perform as claimed constitutes an analysis of real capabilities, which is just as difficult as the more fundamental assessment of the regime's motivations. Nonetheless, their presence cannot be ignored when the GCC calculates its own defensive capabilities and political vulnerabilities vis-à-vis Iran.

Belligerent Posturing

Iran's belligerent claims have escalated as rhetoric has heated. Former Iranian Defense Minister Admiral Ali Shamkhani utilized national television to assert that Iran's much touted Shahab missile – obtained from North Korea in original form – had been fundamentally improved by Iran to produce the Shahab 3 by "strengthening its power of destruction, its range, and its accuracy, and in terms of the capability for local production. We can produce missiles like we produce cars … we have the capability to make missiles like candy."[7] He pointedly contrasted Iran's current independence from "non-Iranian technology" and weapons imports with its predicament during the Iraq war, when only 60mm mortars were available for the crucial battle by the Khorramshahr Bridge— for which missile technology like that of the Shahab 3 was required. Iranian state television followed these comments a week later with the announcement of the test firing of a domestically produced missile able to evade radar and anti-missile missiles. According to Hossein Salami, head of the Revolutionary Guard's air force, "the technology is completely new, without copying any other missile systems that may exist in other countries."[8] Apparently trying to ease regional concerns, General Salami said, "Although the weapons we manufacture are long-range ... they are not supported to increase the concerns of the region's countries. But Iran can block oil export whenever necessary. Both geo-politically and geo-strategically, we control this strategic artery of the world."[9]

Such references to political and strategic Iranian hegemony in the Gulf region, while presumably aimed principally to deter the United States, inevitably resound as pointed hostility toward the GCC. Continuing the same theme of technological innovation and self-sufficiency, the following week Iranian TV aired footage of Iran's, "new, highly-advanced flying boat," already in mass production. According to the Revolutionary Guard's General Mohammad Rahim Dehghani, the flying vessel "cannot be tracked by any naval or aerial radar, and has a velocity exceeding 100 knots per hour, and is capable of carrying and launching different types of weapons."[10] Also displayed during the war games exercise running at the time, according to the broadcast, were two new advanced missiles: the shoulder-launched Misagh 1 – a heat guided surface-to-air missile – and the surface-to-sea Kowsar. Finally, during the same week-long Gulf and Arabian Sea war games in which seventeen thousand Revolutionary Guards reportedly participated, General Ali Fadavi, deputy head of the Revolutionary Guard's navy described their new Iranian-made underwater missile as three to four times faster than a torpedo and able to reach speeds of 223 miles per hour. He said the missile had "a very powerful warhead designed to hit big submarines," and that, "even if enemy warship sensors identify the missile, no warship can escape because of the missile's high speed."[11] From the description, the missile's capability resembles that of the Russian-made VA-111 Shkval developed in 1995.

Although Iran's fledgling space program falls outside this array of touted military achievements, its potential relationship to missile and nuclear development is cause for long-term concern. Consistent with its categorization of its nuclear aspirations, Iran claims its space and rocket efforts are completely peaceful and designed to support telecommunications and the observation of natural disasters. In October 2005, Russia launched on behalf of Iran a camera-equipped micro satellite which orbits the earth every 99 minutes.[12] This modest beginning is apparently merely a prelude to the development of an independent capability to launch its own satellites into orbit with "increasingly large rockets ... in theory the biggest [of which] could hurl not only satellites into space but warheads between continents."[13] The timing and flamboyant publicizing of these developments almost appear designed as deliberate taunts aimed at spooking Washington's hawks into even greater US military visibility in the Gulf, which would demonstrate Iran's defiance and independence.

Altered Gulf Power Balance

While these weapons programs obviously heighten GCC security concerns, it is terrorism and the related risks created by the increasing disintegration of the Iraqi state that generate the greatest anxiety. In simplest terms, Iraq – instead of serving in its traditional role as a regional military balance to Iran – now has its predominantly Shiite southern province significantly influenced if not dominated by the Islamic Republic.

While Saddam Hussein's Iraq exposed the GCC states to serious military threats, its rigid secular Baathist security apparatus created a barrier to jihadist infiltration. Saddam's support of Palestinian terrorism bore the imprimatur of anti-Zionism and Arab nationalist resistance. Although Saddam Hussein – like Iran's theocrats – regarded the GCC rulers as illegitimate, his capacities for subversion were limited and there has been no credible evidence that he assisted Al Qaeda operations or tolerated radical Islamist activities in Iraq.

Iran has direct or proxy assets adjacent to Saudi Arabia's largely Shiite Al-Hasa region in the Eastern Province. With Shiite minorities in many of the other GCC states – and Bahrain with a majority Shiite population – GCC spokesmen have usually muted their concerns about Iran's conventional weapons programs. However, less muting has been evident on the issue of the nuclear program. This is partly due to GCC apprehension over the catastrophic environmental implications of a possible nuclear accident at the Russian-built nuclear reactor at Bushehr in Iran, which is only 150 miles across the Arabian Gulf from Kuwait.

With Iran's notoriously earthquake-prone geology, plus the reportedly Chernobyl-era technology of the Bushehr reactor the possibility of an accident is alarming. Sami Al-Faraj, who heads the Kuwait Center for Strategic Studies, which is advising the GCC on how to prepare for nuclear accidents in Iran, believes "An earthquake could cause an accident that would be more disastrous for Gulf countries than for Iran. A catastrophe that kills 200,000 people could mean wiping out half of Bahrain and pollution of the Gulf would shut down the six water desalination plants on the Arab shore."[14] In a similar vein, Bahraini journalist Sawsan Al-Sha'er commented on Al-Arabiya TV that, "My problem is that the nuclear reactor is closer to Manama than to Tehran. Even if it were just an environmental issue – forget about security – I would still have a problem with it."[15]

The Nuclear Double Standard

GCC environmental concerns can be expressed within a far less controversial Arab-wide framework than the nuclear weapons issue, which runs head on into the "double standard" issue. This, of course, is the long standing and frequently repeated Arab objection to US acceptance of Israel's large, highly developed nuclear arsenal. With a full range of weapons delivery systems, as well as mid-air refueling and submarine launched second strike capabilities, Israel's regional nuclear hegemony remains a source of major controversy for US and European–Arab relations. The US confrontation with Iran has brought this controversy to the forefront and complicated the GCC's publicly expressed case against Iran's nuclear program. Objection to Iran's program means joining a Western campaign; a campaign which is generally silent regarding Israel's nuclear arsenal. In January 2006, the Secretary-General of the Arab League, the Egyptian Amr Moussa, sent a message to the leaders of the GCC summit asking them to focus on Israel rather than Iran. He repeated this request at a later Arab meeting saying, "We should avoid double standards."[16]

Ambivalent and at times contradictory views on the issue have marked the unofficial Egyptian reaction. For instance, the Muslim Brotherhood in Egypt, notable for its recent stunning electoral result in Egypt's parliamentary elections, has spoken with several voices on the subject. Essam el-Erian, a Brotherhood spokesman, said, "If Iran developed a nuclear power, then it is a big disaster because it already supports Hezbollah in Lebanon, Hamas in Palestine, Syria and Iraq, then what is left? We would have a Shiite crescent that the Jordanian King warned against."[17] Two months later, however, deputy Brotherhood leader Mohammed Habib stated that the Brotherhood, now Egypt's largest opposition party, believed Iranian nuclear arms "would create a kind of equilibrium between the two sides—the Arab and Islamic side on one side and Israel on the other."[18]

GCC reaction, however, has for the most part been increasingly negative. In December 2005, the Emirates Center for Strategic Studies and Research (ECSSR) called on the GCC to avoid silence on the nuclear issue, as the Gulf States would pay the price for escalation between Iran and the West.[19] Saudi Arabian analyst Dawood Al-Shirian has asserted that, "Gulf nations utterly refuse any idea that Iran should own a nuclear weapon, and they want Iran to stop uranium enrichment … except under international control."[20] Former Kuwaiti Minister Dr. Ahmad Al-Rubei, interviewed on Al-Rai TV, dismissed the logic that because Israel has nuclear bombs, "we" must also have them, "What we are facing is madness. The Iranian nuclear activities must cease. If you drop a bomb on Israel, what would you say to it? … strike the Jews and spare the non-Jews … the million Palestinians in Israel and what about the neighbors in the region, the people of the West Bank and Gaza, Amman and Damascus? and [referring to an Iranian official's reassurance vis-à-vis Bushehr's Chernobyl-era technology] 'Don't worry, this reactor operates not on Russian technology, but on pure Iranian technology.' What pure Iranian technology? We fear the catastrophes of the Russian technology, and you're talking about pure Iranian technology?"[21]

The Proliferation Enigma

Possible nuclear proliferation resulting from Iran's program is discussed on the sidelines of the military and environmental issues demanding GCC attention. Saudi Arabia, considered the most likely next nuclear developer, has categorically denied any such intentions. However, it would appear that, at least over time, a nuclear-armed Shiite Persian Iran would constitute an unacceptable challenge to the Sunni Arab guardians of Mecca and Medina. With ample financial resources and a historically close military relationship with Pakistan, Saudi Arabia would presumably be able to arm itself with nuclear weapons short of a lengthy and inevitably visible technological development process.

The Saudis proved themselves capable of successful deception when they secretly purchased between 35 and 60 CSS-2 intermediate-range surface-to-surface ballistic missiles from China, based on negotiations that began in 1985 which also covered mobile launchers and technical support.[22] The story did not break until March 1988.[23] Washington was alarmed because the missiles had been transported among Chinese military supplies destined for Iraq and then hidden in the Saudi Rub al-Khali (Empty Quarter) where they remained unknown to the US for two and a half years. Nuclear-related suspicions arose because the missiles had been withdrawn from a Chinese operational nuclear missile unit and were considered unsuitable for conventional weapons because of their inaccuracy. Their range was between 1,550 and 1,950 nautical miles, putting all of Saudi Arabia's neighbors in reach—including Israel, which responded by simulating air attacks on Saudi missile facilities. The uproar created among the Kingdom's neighbors and the international anti-proliferation community, as well its implications for the US–Saudi relationship must have caused Riyadh to doubt whether the acquisition was worthwhile. Although the Saudis have refused to allow the United States to inspect the CSS-2's, once the initial furor had subsided they signed the Nuclear Nonproliferation Treaty (NPT).

Although in the final analysis the Kingdom might well opt for a nuclear capability if sufficiently threatened by Iran, the decision would involve conflict between the desire for secrecy – to avoid extremely adverse US, Israeli and other international responses – and the need for public disclosure not only to deter Iran, but to bolster the Saudi defense posture and prestige in the Muslim world. Western fears over Riyadh's nuclear status would mount as the uncertainties of a generational succession process in the royal family approached. Western–Saudi planning must also consider the possibility of an Egyptian nuclear program eventually emerging to create additional pressure. But like the Saudi situation, Egypt's intentions remain in the realm of rumor and speculation.

GCC Defenses

Despite Iran's threats and recent posturing of advanced weapons during exercises in the Gulf, Tehran's claims to dominate the Gulf are overblown. The GCC states have steadily developed their own defense capabilities. These are considerable in terms of fighter aircraft and missile defense, but overall GCC defense planning has traditionally been hampered by a lack of interoperability. Rather than devising an overall air, sea and ground defense capability, each state has chosen to go its own way. Coordinated defense planning, though discussed by the GCC for many years, has mostly resulted in brief military exercises with no integrated follow-up.[24]

Constructive ideas and plans for a critically important joint air defense system were commissioned in 2001. Such a system would tie in with the GCC air defense structure, allowing GCC states to track aircraft in their airspace and coordinate responses. Brig. General Khalid Al Bu Ainnain of the UAE Air Force, has proposed the development of S-band radars on three fronts – northern Saudi Arabia, the UAE and Oman – to enhance warning of missile attacks.[25] Even with many of these improvements either incomplete or set aside, the Saudis and the UAE have significant air forces, air defense systems and naval assets, particularly when supplemented with other Gulf state assets. The GCC is therefore in no way incapable of inflicting a damaging response to any conventional Iranian attack.

As noted above, however, the GCC's vulnerability vis-à-vis Iran, irrespective of nuclear weapons development, arises from the clerical regime's newly enhanced ability, largely due to the situation in Iraq, to stir up trouble among Shiite minority populations in the GCC. Should such destabilization efforts escalate to the stage of asymmetric warfare the GCC would be at a grave disadvantage. Al Qaeda-sponsored terrorism, against Saudi Arabia in particular, already poses a significant terrorist threat, as evidenced by a carefully planned attack against a major oil facility at Abqaiq in February 2006. Although thwarted and the perpetrators reportedly captured, the effort provided tangible evidence that Al Qaeda's threats against the Saudi oil structure had moved from rhetoric to implementation.

To date, Al Qaeda and terrorist organizations connected to Iran have pursued some overlapping objectives – in Lebanon and Palestine/Israel – but aside from Hezbollah attacks on American installations in the Kingdom, Iran apparently has refrained from covert actions in Saudi Arabia. With American forces no longer in Saudi Arabia Iran has less incentive, at least at the moment, to mount risky operations on the peninsula. Al Qaeda's stirring of the Sunni insurgency in Iraq through Abu Musab Al-Zarkawi's terror operations worked against Iran's ambitions for a Shiite-dominated Baghdad national government or at least a stable Shiite southern Iraq able to prosper and expand Iran's influence. Therefore, the GCC focus for now will likely continue to be on thwarting Al Qaeda terrorism, which may reach much more dangerous levels should Iraq's turmoil worsen.

Implications of Iraq's Upheaval

The evolution of events in Iraq will have greater impact on the security of the GCC states than almost any other factor. This is true not merely because of the sectarian tensions now stimulated but because of the implications of a perceived American military defeat.

Jihadists throughout the Islamic world would trumpet the defeat of a great superpower and Islamist political parties in the Arab world would gain significant momentum as they capitalized on the dramatization of an Islamic victory over "crusader" legions. Should sectarian fighting intensify to full, organized civil war, the entire region will undoubtedly become involved. Already exacerbated tensions would worsen as Sunni states reacted to the likely presence of Iranian forces actively engaged in Iraq to solidify the Shiite majority position.

Possible scenarios are numerous and many look detrimental to the stability of the Arabian Peninsula. The possibility of partition in Iraq is real. According to a reformist member of Iraq's parliament, Sayyed Ayad Jamal Al-Din, a religious scholar interviewed on March 26, 2006, "There is a widespread on-going and unabated expulsion of Shiite families from [Sunni] areas such as abu Ghraib, al-Tarmiya and al-Taji. So far, 3,700 Shiite families were expelled, in addition to car bombs that target the working people who are Shiite."[26] A serious expansion of this action brings to mind the humanitarian catastrophe following the partition of the Indian subcontinent after independence in 1947. Many tens of thousands were slaughtered as they fled toward a communal safe haven.

Saudi Arabia has watched the Iraqi population in Jordan swell to an estimated 800,000 due as much to communal violence as to the battle against the US occupation per se. The Kingdom reportedly has invited bids for the construction of a security fence to span its 560 mile border with Iraq.[27] Whether such a project would stop determined Iraqi escapees of chaos or partitioning seems uncertain. Also, such a fence could not be counted upon to exclude Saudi or other jihadists returning from Iraq who would probably find Syria a more hospitable route. The Kingdom's apparent progress in containing in-country terrorist attacks could be seriously set back by an influx of jihadists experienced in bomb making and covert operations. However, if such an influx were symptomatic of an Iraq

– or a newly federated part of Iraq – bent on eliminating terrorists and preventing cross border operations against GCC states, the temporary additional threat might prove acceptable.

Any analysis of GCC security in relation to Iran's nuclear program must therefore weigh the potential consequences of a US or Israeli military strike on Iran; the regional impact of events in Iraq as efforts to form a national government either succeed or fail; and how best to deal with an Iran bent on politically exploiting the possession of nuclear weapons—whether actual or merely prospective. Simultaneously the GCC – and particularly Saudi Arabia – must continue to estimate the impact of its responses as a major target for Al Qaeda and its affiliates. As is the natural tendency to prefer a crisis some years away to one that might occur tomorrow, there are many indications that most GCC analysts would choose to deal with a future nuclear Iran over the inevitable military and political shockwaves caused by an attack on its nuclear installations.

Weighing the Preemptive Option

Although the GCC states certainly have their own intelligence assets in Iran, their means of judging a timeframe for Iran's program is probably as clouded as that of the West. Intelligence forecasting in the region has earned itself a bad name. Iraq was able to conceal its nuclear intentions in spite of UN inspections until the defection of Saddam Hussein's son-in-law to Jordan. The confusion over Iraq's weapons of mass destruction, judged by the combined assets of the West and Israel, still remains a source of tension and argument. However, Iran's protestations of peaceful intent must be weighed not only against IAEA inspections, but against the recent history of both India and Pakistan's successful covert nuclear weapons development. Estimates of when Iran might produce a bomb vary from ten or twenty years, to two years or within months if components were imported. The possibility of US military action on the other hand is conceivably imminent; an Israeli strike seems much less so. If the publicly available assessments are to be believed, Israel, which has had ample provocation from Iranian President Ahmadinejad, could not by itself easily reach Iran's numerous dispersed nuclear development installations. Distance, coupled with the number of sorties required, would stretch Israel's assets unreasonably.

A US attack – unfortunately in public minds potentially confused with Israel's one shot destruction of Iraq's Osiraq facility in 1981 – would require a major exercise of war. With the precise location, degree of hardening and depth underground of Iran's dispersed facilities, an attack would require many sorties and would have to be preceded by extensive repression of Iran's air defenses, including sophisticated Russian and North Korean surface to air missiles, anti-aircraft guns, fighter aircraft and command and control facilities. One former head of the US Defense Intelligence Agency's Middle East section predicts a US assault would require "in the neighborhood of a thousand strike sorties [with] all kinds of stuff – air, cruise missiles, multiple re-strikes – to make sure you've got it all."[28] Such attacks would occur over a number of days and would have to include a concentrated campaign to preempt Iran's undoubted reprisal efforts that are certain to be aimed at GCC allies of the United States as well as US targets in the Gulf itself. An Oxford Research Group study published in February 2006 concluded that a US attack aimed at setting back Iran's nuclear program by at least five years would also "require destruction of Revolutionary Guard facilities close to Iraq and of irregular naval forces that could disrupt Gulf oil or transit routes."[29]

A preemptive US attack on Iran would therefore present the GCC with a devastating scenario. Given no option but to sustain its basic defense relationship with the US attacker, the Arab Gulf states would be exposed to critical physical as well as political damage. Even if Iran's immediate reprisals against US military facilities – particularly in Bahrain, Qatar and the UAE – were minimal, an extended Iranian reprisal effort could be anticipated. This reflects the core problem—that the extensive and lengthy US military action required for the operation would ultimately constitute a war between Iran and the United States. Such a war would draw US military forces ever more deeply into the region, fortifying the extremist anti-western ideologues and the jihadists themselves. These consequences would not be confined to the immediate Gulf region but would extend throughout a Muslim world already profoundly divided in its efforts to reconcile the pulls of modernity with traditionalist demands.

Iran's retaliation could extend to the disruption of GCC oil exports as a means of punishment that would also hurt the Western economies by raising oil prices to astronomical levels. This temporary benefit to the clerical regime, however, could escalate the conflict toward "the vast majority of Iran's crude oil reserves ... located in giant onshore fields in the southwestern Khuzestan region near the Iraqi border."[30] Several bombs were detonated near Khuzestan oil wells in September 2005 – suggesting vulnerability – which could be attributable to ethnic Arabs of the region or even to US or British Special Forces operations designed as signals to Tehran.

US and Israeli Reactions

As though echoing among the deaf, these dire assessments – along with Iran's reckless threats and the urgent demands for drastic action against Iran from many quarters in Israel and the US – all resonate without mentioning adverse consequences. An impassioned *Wall Street Journal* editorial in February 2006 foretold that Iran would "use the leverage of the bomb to dominate the Middle East and limit America's ability to defend itself and fight terrorism [and be] the gravest threat in the world to U.S. interests." The editorial quoted Ahmadinejad's outrageous threats against Israel as well as Ali Akhbar Rafsanjani's [now head of Iran's powerful Expediency Council] earlier bluster that a nuclear bomb against Israel "would leave nothing on the ground, whereas it would only damage the world of Islam." He was perhaps borrowing from a former Indian prime minister's comment that while nuclear exchanges would destroy several Indian cities, they would eliminate Pakistan as a nation. The *Journal* predicted the bomb would allow Iran to dominate OPEC, disrupt maritime traffic in the Gulf, "and force the U.S. navy out of its shallow waters … menacing Europe and eventually the U.S. mainland."[31]

Unsurprisingly, a similar reaction sprang from the main elements of the Israeli lobby in the US during late 2005, where the American Israel Public Affairs Committee (AIPAC) relentlessly criticized the Bush administration for its patience in supporting efforts for a diplomatic solution. The AIPAC described Bush's efforts on Iran as "dangerous," "disturbing" and "inappropriate."[32] In the past, the AIPAC has occasionally attempted to steer US policy even beyond the wishes of the Israeli government at the time. In the case of Iran, however, any pressure for US restraint is unlikely except from a minority of analytical Israelis prepared to measure the comparative consequences of giving up their nuclear monopoly in the Middle East as opposed to coping with the multitude of long-range Iranian retaliations to be endured following a US – or Israeli – attack on Iran.

The explicit Iranian presidential threat of nuclear annihilation has vitiated all prospective Israeli public arguments for acceptance of a nuclear-armed Iran. In the absence of a significant change of leadership in Tehran or a heroic and tightly verifiable diplomatic success by the IAEA, the GCC must expect the impact on US politics to remain profoundly influential. This is so not only in relation to concern for Israeli security but also for US interests in the Gulf and beyond.

Viewed from the GCC there are many ironies to this perfect storm of deadlocked confrontation. The United States branded Iran a member of the "axis of evil," then removed the Taliban government from Afghanistan, in part with Tehran's cooperation. The destruction of Iran's traditional Baathist enemy in Iraq followed, thereby increasing Iran's regional security and relative military power to an unprecedented level. The tradeoff for Iran, however, is the presence of American forces in Afghanistan, Pakistan and Iraq, which grates painfully on long-standing paranoia about US intentions. When coupled with "regime change" threats substantiated by funding for the Iranian Mujahedin-e-Khalq (MEK) opposition group and US information programs designed to separate the clerical leadership from popular support, Iran's quest for a nuclear deterrent can only be quickened.

Iranian Judgments

There is every indication, however, that rational analysis devoid of religious zealotry survives in some quarters of Tehran's policy apparatus. If so, it will address the question of whether the possession of nuclear weapons would really offer immunity from a US attack and add significantly to their prestige and ability to influence and manipulate regional events. Such an Iranian analysis would also need to examine Iran's basic vulnerabilities. If Iran were to reveal actual nuclear weapons and at the same time display credible missile delivery systems – whether actual or under development – could it expect a benign international reaction, particularly from the United States? If Iran's defense posture was based on missile defense, how much of this capacity would remain following a US air and cruise missile campaign accompanied by strikes against nuclear facilities? With subsequent US control of Iran's air space, how would the Republic protect its remaining assets and present a sufficiently credible military posture to match its rhetorical boasts? Iran might be reduced to stepping up terrorist attacks through Hezbollah or its networks operating from embassies abroad. These attacks would all have a home address, however, guaranteeing devastating retaliation against Iran.

Is Iran's economy strong enough to satisfy its public's expectations? especially when, "Overall, Iran's oil sector is considered old and inefficient, needing thorough revamping, advanced technology and foreign investment."[33] Oil production is far below pre-revolutionary levels and requires billions in foreign investment to keep pace with population growth and industrialization. The country now produces about 4.2 mbpd, but exports only 2.7 mbpd due to a domestic consumption of approximately 1.5 mbpd. As it is, Iran spends about $4 billion on gasoline imports, amounting to one third of annual consumption.[34] With a fourteen percent unemployment rate, which is much higher among the younger population, Iran does not look like a country prepared for a long-term heightened confrontation with the West. However, it talks like one.

In political terms, the clerical hardliners cannot overlook a recent Ministry of Guidance and Culture survey which indicates only 14 percent of the population attend Friday prayer services. Also, despite the nuclear fervor that has brought an emotional surge of nationalist unity, Iran's political divisions remain. Beyond this, the country's ethnic minorities constitute a potential vulnerability that could be exploited by foreign intervention aimed at destabilizing Iran. With only 51 percent of the population considered ethnic Persians who speak Farsi, the rest of the Islamic Republic is mostly divided among 24 percent Azeri, 8 percent Gilaki and Mazandarani, 7 percent Kurd, 3 percent Arab, 2 percent Baloch and 2 percent Turkmen. Among these, the Kurds are probably the predominant concern as the quasi independence of Iraqi Kurdistan inevitably exerts strong cross-border magnetism. Iran's awareness of its separatist vulnerability was recently acknowledged by the Revolutionary Guard's theoretician Hassan Abassi during an interview with a reformist Iranian internet daily in which he stated, "America's program today is to dismantle Iran into seven countries, and to exploit its ethnic disputes."[35]

Iranian Pragmatism?

Trying to apply analytical logic to Iran's decision-making process is tantamount to grasping at straws. There have, however, been instances when the clerical regime has subordinated its fervent Islamic orientation to pragmatism. Iran has refused to support the Islamists leading the Chechen separatist rebellion and has sided with Christian Armenia in its long struggle with Azerbaijan over the Nagorno-Karabakh enclave. Nonetheless, the GCC's security analysis must assume that Iran's present leaders are not likely to bend from continued nuclear pursuit.

Ray Takeyh, one of the most astute interpreters of the Republic's revolutionary culture, believes that Ahmadinejad's electoral victory represents the ascendance of an ascetic "war generation." Ayatollah Khomeini's bitter acceptance of a truce, ending the war after eight years of sacrifice, meant betrayal and a sacrilege of unfulfilled objectives to this generation of the "armies of God"; it also betrayed the revolution itself. Seen as a war between secular Baathism and Islam, surrender to Saddam Hussein represented an ideological defeat, attributable to the West's assistance to Iraq and acceptance of Baghdad's use of the chemical weapons that ultimately overwhelmed

Iran's forces. It is this Western role which Takeyh believes accounts for Ahmadinejad and his group of hardliners' implacable distrust and suspicion of the US and the West in general.[36]

Takeyh suggests there is little prospect of compromise between the West and Tehran, at least during this heated phase of confrontation. Beyond efforts to improve missile defenses and better coordinate air surveillance, the GCC appears to be facing yet another Gulf war that is beyond its capacity to influence or prevent.

Therefore, in viewing regional prospects GCC leaders will focus on Iraq, after exhausting opportunities to exercise their own calming diplomacy on Iran. Because terrorism tops the list of GCC security concerns, the direction of events in Iraq will profoundly affect the rest of the Gulf. This issue is closely related to that of a nuclear Iran. Should diplomacy at least temporarily deescalate the US–Iran confrontation, possibilities for cooperation between the two countries to stabilize Iraq might materialize if Iraq were to slip to a level of chaos that Iran found threatening to its own security. This could at least postpone a confrontational crisis. Alternatively, events in Iraq could evolve to worsen the confrontation over the nuclear issue and add further serious dangers to GCC security. The region's future clearly hinges on whether Iraq remains a unified state; becomes a grouping of entities within a federated state; or descends into a chaotic civil war, inevitably involving neighboring states.

Iraq's Disintegration: Effects on the Region

Each of Iraq's neighboring states would become embroiled in such a scenario:
- Turkey to subdue the dynamics of an independent Kurdish state in northern Iraq that would inspire demands for succession among its historically restless Kurdish minority;

- Syria to assist as a conduit for Iran's support of Hezbollah in Lebanon, possibly suffering destabilization effects as its predominantly Sunni population tried to assist their Iraqi brothers;
- Lebanon to maintain its fragile balance as Hezbollah – with historic ties to Iraqi Shiites and the only armed militia remaining after Lebanon's civil war – gains power with Iran's new reach;
- Jordan, as Iran-supported Hamas – already powerful in Palestinian refugee camps and flushed with victory in Palestine – combines as a threat with the followers of Jordanian-born Abu Musab Al-Zarkawi, the former leader of Al Qaeda in Iraq who had conducted terror operations in Jordan;
- Saudi Arabia, as already noted, out of concern for the loyalty of its Shiite Eastern Province;
- Kuwait with an important Shiite minority that would have at least political support from across the border; and
- Bahrain with a Shiite majority population uneasily ruled by a Sunni monarchy subject to terrorism and political destabilization.

The extent to which the United States can manipulate the outcome remains at the center of debate. Even at the extreme limits of the US public's acceptance of casualties and financial costs in Iraq, an outcome approximating victory as envisioned by President Bush remains an unrealistic (even medium-term) expectation of the strategy he outlined in a report in December 2005. Victory was defined as the establishment of "a new Iraq with a constitutional, representative government that respects civil rights and has security forces sufficient to maintain domestic order and keep Iraq from becoming a safe haven for terrorists."[37] Victory in Iraq would also mean a self supporting free-market economy. Even if the insurgency is no longer able to cripple Iraq's oil infrastructure, the process of rebuilding and modernization will extend for decades according to current assessments.

Even in a best-case scenario, US forces will remain heavily in support of Iraq's military and police. Although a government that is reasonably acceptable to the country's factions might stabilize in Baghdad during the next few years, US logistic support would be required until Iraqi units were self-sufficient. With virtually no operational Iraqi fixed-wing or helicopter assets, the United States would remain responsible for border defense and air surveillance for several years. The problems of developing military and security competency are already evident. Militias need to be disbanded and absorbed into national units with no sectarian or ethnic coloration. The US Army has been engaged in a similar endeavor in Lebanon for decades but with dubious expectations for success should violence erupt again. Furthermore, there will be an inherent conflict between building well-led, motivated and competent forces without creating power centers able to challenge or connive with an ambitious faction of the civilian leadership. Throughout, the priority of the US and certainly the GCC will be to prevent Al Qaeda from developing a stronghold in the heart of Sunni Iraq from which to operate against the entire region. The GCC will also be faced with a continuing, large US military presence in Iraq, Kuwait and probably Bahrain, Qatar and the UAE, just as fundamentalist and nationalistic aspirations begin to push increasingly against a Western military presence. Unfortunately the eruption of sectarian carnage in Iraq, occasioned in the name of the quest for democracy, has weakened reformist movements in the region and throughout the entire Muslim world.

Conclusion

For many months to come, Arab Gulf leaders will be subject to increased levels of tension. The judgments and wisdom of Iran's clerical leadership have been shown to be reckless and unsound. They have threateningly challenged a superpower led by hawks with substantial ground forces on two of their borders, major close-by air bases on all sides and significant naval forces within reach. They have also urged the destruction of Israel, the region's military hegemon and by denying the most thoroughly documented genocide in history – the holocaust – they display a desperate reach for greater popularity, disdainful of the taunt's powerful meaning to Jews.

While the threats to GCC security are all too real, their capacity for damage is easily overlooked in the heat of the confrontation and ongoing crisis in Iraq. Iran capitulating to the West's demand to drop its nuclear development now appears as impossible as the West's acceptance of this development, but measured against the losses each side will endure in the absence of compromise, the possibility of some form of solution becomes more likely. Iran's clerics have far reaching goals, but do these require nuclear weapons once the leadership can be convinced that immediate military threats to their survival are removed? The clerics presumably see themselves seizing the moment by asserting leadership in an Arab world lacking heroes, disillusioned by the lingering debris of unsuccessful governing philosophies; the uniting promises of Arab nationalism, Baathism, communism and the failed parliamentary systems inherited from European powers. Could Iran's fervor capitalize on this malaise to inspire Islamists to replace Western-influenced governments? Is there sufficient evidence that an Iran militarily enhanced by nuclear weapons would gain more capacity to lead and inspire? Beyond these questions one outcome of Iran's belligerent posture is evident—while the clerics may fancy themselves as the vanguard of a final expulsion of Western domination and military interference, they have simply invited an enhanced, long-term role for Western forces in their region.

For the realist elements in Iran, such ambitions are surely tempered by the realities of the situation. If assessments are correct that Iran's government is unpopular with 80 percent of its population, how charismatic an image does the Republic really project?

Nuclear issue aside, the GCC will surely examine with a skeptical eye the real but easily exaggerated issues of ethnic and sectarian separatism that have been repeatedly raised throughout this study. The forces of the "Shia crescent" propelled by events in Iraq will at some point become over-identified with Iran and resistance will be encountered. For instance, Hezbollah in Lebanon is being accused of loyalty to Iran and Syria over Lebanon itself. The divides of Shia versus Sunni; Arab versus Persian; Kurdish Sunni versus Arab Sunni or Shia; or sub-sect against sub-sect within faiths are intermingled with national identities. The more alarmist views of Iran's new power emphasize its Shia magnetism as a potential instrument for the Islamic Republic to bend other states to its will. However, at some point – perhaps not until serious damage has been done – Shia-inspired actions will become identified as Iranian interference, swinging the pendulum back toward nationalism. The GCC states may conclude that even with a nuclear Iran on their doorstep they will be able to retain the loyalties of their Shia minorities.

Finally, as the proliferation issue has been moved to the forefront of international politics by confrontation in the Gulf, new opportunities arise for the GCC to lobby for a nuclear-free Middle East. Although the problem of devising a verification system which is convincing to Israel appears insurmountable at the moment, it is not beyond the capacity of the world's scientific community. Nor is the diplomatic challenge so insurmountable now that the alternative has become so starkly apparent.

4

Israel and the Strategic Implications of an Iranian Nuclear Weapons Option

Geoffrey Aronson

There has long been a strategic dimension to relations between Israel and Iran. For many years Israel viewed Iran as part of the outer circle of nations bordering on the Arab world which shared a mutual concern about the development of Arab military power. This provided the foundation for wide-ranging cooperation during the years before the Iranian revolution, including on the funding and development of missile delivery systems that could be adapted to nuclear use.

The "strategy of the periphery" was a doctrine authored by Israel's first Prime Minister, David Ben-Gurion, in the years after Israel's creation. The idea motivating this strategy was disarmingly simple—Israel, its proponents argued, shared a natural communal interest with the other non-Arab (and frequently non-Muslim) states and minority peoples (Maronite Christians, Kurds) in the region. Beginning in the 1950s, Israel established wide-ranging covert alliances with regimes in Ethiopia, Turkey, Lebanon and Iran—each of which was enmeshed in contests with its Arab neighbors. Relations with the Maronite leadership in Lebanon were based on mutual antipathy towards pan-Arabism. In addition, military ties were established with Iraqi Kurds, who under the patronage of the

Shah mounted an insurrection against Baghdad that only ended when Tehran withdrew its sponsorship of the rebellion in 1974.

During the Shah's reign, Israel and Iran enjoyed close relations. One of the most noteworthy and strategically significant collaborations was "Operation Flower." By 1977 Israel and Iran had reached a secret agreement on the joint development of Israel's Jericho II intermediate range ballistic missile. According to the terms of Operation Flower, Iran paid Israel the equivalent of $260 million in oil and permitted test flights over Iranian territory. However, the fall of the Shah put an end to the joint program.[1]

The aim of these alliances of the periphery – which despite Israeli entreaties remained clandestine and limited largely to military affairs – was to contain the power of Israel's Arab antagonists and to divert their resources away from the struggle against Israel.

The success of the Islamic revolution in Iran struck a critical blow to this strategy. At the height of the campaign against the Pahlavi throne in 1979, Ariel Sharon – then minister of agriculture – and others proposed that Israeli paratroopers be dispatched to Tehran to crush the revolution and restore the Shah. Yet the shared hostility towards Arab power, and particularly the pretensions of Saddam Hussein's Iraq to develop a non-conventional nuclear option, continued after the Shah's fall. Indeed, Iranian planes targeted Osiraq before Israel's far more successful operation against the Iraqi nuclear reactor in 1981.

Even after the revolution, Israel still hoped to come to some modus vivendi with the mullahs and continued its efforts to forge ties with the Islamic Republic—providing arms for the war against Iraq and encouraging Washington's famous tilt against Baghdad through the Iran Contra scheme during the Reagan years.[2]

Despite the renewal of clandestine military relations during the Iran–Iraq War, relations never again included issues related to weapons of mass destruction or their delivery systems. Indeed, it became clear that the nature of the Iranian regime itself, rather than any other consideration, had become the critical factor in determining Israel's hostile attitude toward Iran's post-revolution development of its nuclear capabilities.

Israeli officials and analysts base their antipathy towards Tehran on a number judgments: First and foremost, Iran is viewed as a competitor for influence and power in the region – challenging the creation of a post-Cold War network of pro-American and client regimes led by Israel – which seeks to undermine an Israeli vision of the region that assumes Israel's preeminence and its nuclear monopoly. As Prime Minister in the early 1990s, Yitzhak Rabin decried Iran's "megalomaniac tendencies of becoming the Middle East superpower, while exploiting radical Islam in all its aspects to destabilize Arab regimes."[3] Rabin's tenure established the policy and conceptual framework that guides Israeli policymakers to this day. He believed that Tehran was the source of "all the threats by all types of fundamentalist Islam" and that Iran, with the cooperation of European and Chinese suppliers, had embarked upon the development of weapons of mass destruction, including chemical, nuclear and biological capabilities, which posed a far greater long-term threat than Syria.

The period before the development of an Iranian non-conventional capability would offer Israel a "window of opportunity" to isolate Iran by stabilizing the Arab–Israeli conflict and creating an Arab–Israeli community of interests which, allied to Washington, could frustrate Iranian aims and postpone any nuclear "day of reckoning." "Our problem," noted Rabin in late 1993, "is how soon can we reach peace agreements or the neutralization of hostilities with the immediate circle around us, before the threat from the more distant circle intensifies."[4] This Iranian dimension of the now discredited Arab–Israeli peace process was a rarely acknowledged but nonetheless critical element of Israel's strategic outlook during the era of the abortive Madrid diplomatic process.

General [ret.] Ephriam Sneh, a Rabin confidant, established himself over a decade ago as one of the most outspoken policymakers on this issue. "If," he noted, "despite all our precautions, we are confronted with an Iran already in possession of nuclear installations and in mastery of launching techniques, we would be better off if the explosive charge of the Arab–Israeli conflict is by then already neutralized by peace treaties with Syria, Jordan and the Palestinians."[5]

Like his predecessor Yitzhak Rabin, Shimon Peres was comfortable demonizing the regime in Tehran, which Rabin liked to describe as "Khomeniism without Khomeni." In a speech before the National Press Club in Washington, Prime Minister Peres offered his strategic rationale for concluding peace between Israel and the Arab world: "We have to try hard and seriously to get ourselves organized against the new cloud which is endangering the skies of the Middle East – fundamentalism – with Iran serving as the headquarters, with an attempt to get non-conventional weapons, with the use of terror all over the place ... Iran is the greatest danger to the Arabs, to the Israelis, to the peace in the whole of the Middle East and outside the Middle East. They are cheating. They are lying. They are conducting a murderous campaign all over the place. They are financing, training and commanding Hamas and Jihad ... but still they are trying to get hold of a nuclear option. And the combination of an evil wind with non-conventional arms poses the greatest danger to all of us."[6]

Soon after PLO leader Yasser Arafat's arrival in Palestine in mid-1994, US President Bill Clinton's National Security Advisor Tony Lake made an important speech at the pro-Israeli Washington Institute for Near East Policy. Lake made clear the connection US policy makers had drawn between the Gulf War, Oslo and the expansion of American power against the two principal "rogue states" in the region—the secular regime of Saddam Hussein in Iraq and the Islamic radicalism of the mullahs in Tehran.

Lake explained to his audience that the peace between Israel and its Arab neighbors was meant to be an aggressive peace, aimed at isolating the "backlash states" headquartered in Baghdad and Tehran. Saddam Hussein would continue to be ostracized by Arab regimes joining a US-led coalition of peace and economic development. Iran would be deprived of its strategic relationship with Syria, delivering a double blow against terror and the prospect of a non-conventional war. Borrowing an idea favored by Israeli Prime Minister Rabin and Foreign Minister Peres, such a peace would also constitute the realization of an Arab–Israeli coalition against Islamic extremism, which was defined by Lake as "a threat to our nation's interests."

Lake stressed that with the stakes so high the US could not afford to be a bystander. Indeed, Lake declared that US policy in the region was "a paradigm for our nation's approach to the post-Cold war era." In the struggle between "violence or peace, regression or freedom, isolation or dialogue," the power and prestige of the United States, Lake promised his audience, would be engaged.[7]

Clearly, the "strategy of the periphery" was ill-suited to this emerging era of rapprochement with the Arabs. The campaign for Arab–Israeli peace, on Rabin's terms, was the centerpiece of Israel's effort to contain Iran and to forge an Israeli–Arab consensus supporting this objective. Rabin worked with great, if not complete success to convince Washington – where the theory of "dual containment" then ruled – of the need to undertake similar efforts across a broader front, including pressuring Europe, and notably Germany and Switzerland, as well as Japan and China to minimize their lucrative economic and military relations with Tehran.

Iran was on Rabin's mind on November 16, 1993 when he was asked to state Israel's policy on the use of nuclear weapons: "Israeli policy was and is that we will not be the first to introduce nuclear weapons to the context of the Arab–Israel, Islamic–Israeli conflict." Rabin thus provided explicit warning that Israel's strategic nuclear horizons had spread unambiguously to Tehran.[8]

The Israeli Campaign Against Iran

For more than a decade, Israel has been at the forefront of those elements of the international community that express concern about Iran's ongoing nuclear-related programs. Israel has attempted to raise the international profile of such efforts and to spur international action against Iran's nuclear activities. On June 16, 1992, David Levy – then foreign minister – officially informed the United States of Israel's concern about Iran's nuclear pretensions. Israeli reports noted that Western firms, some of which formerly assisted Iraq, were assisting Iran and Libya.

The Rabin government viewed US efforts to preempt Arab and Iranian non-conventional capabilities as the litmus test of its support for Israel. "This is the test of the United States in the region," noted Yitzhak Rabin. "Whether it will manage to curb and prevent the proliferation of nuclear weapons … Only the US can do this and I hope that it will," he said. Indeed, Rabin declared that, "the importance of our relationship with the United States should be seen in this context."[9]

The Madrid process was understood by its proponents in Washington and Jerusalem as a vehicle for isolating Iran. "For the first time since 1948, a new existential enemy has been named," noted Israel's leading newspaper *Ha'aretz*. "Huge armies of the neighboring countries now look much less dangerous than those of a distant state which seeks hegemony over the entire Middle East with the help of nuclear power, terrorism and Islamic fundamentalism."[10]

Iran opposed the emerging, post-war structure of relations being created under American leadership, symbolized by the Madrid diplomatic framework. Not surprisingly, it viewed this effort as a threat to its own ambitions and security. As Rabin's chief of staff Ehud Barak noted, "Iran would still build up its army and attempt to attain nuclear power even if Israel did not exist in the region. Iran uses Israel as an emblem for its struggle, but since it is threatening the [Arabian] Gulf, the free flow of oil and the stability of more pragmatic regimes surrounding them, I believe that the international community, under US leadership, should deal with the Iranian challenge."[11]

Israel labored to convince the United States to take the leading role against Iranian nuclear pretensions, supplementing the aggressive peace pursued under the Madrid banner. Israeli experts believed that Iran's entry into the nuclear club could be preempted. Ephriam Sneh told a symposium that it was "still possible to prevent Iran from developing a nuclear bomb. This can be accomplished since Iran threatens the interests of all rational states in the Middle East … If the Western states do not fulfill their duty, Israel will find itself forced to act alone and will accomplish its task by any means suitable for the purpose … During the coming stage of negotiations, Syria's links with Iran must be questioned so as to force Syria to declare to which of the two camps it belongs," he said. Israel expected the United States, in the words of then Foreign Minister Shimon Peres, "rather than us [to] stand at the head of this campaign" against Iran.[12]

During the 1990s, however, military preemption remained an implicit option rather than an operational strategy. Israel continued to feel confident of its deterrent power, even vis-à-vis a nuclear-armed Iran. While acknowledging that, "there is no doubt that such a [Iranian non-conventional] threat is beginning to loom," Chief of Staff Matan Vilnai explained, "we will not invade Iran or declare war on it, but we have to create conditions which will make Iran think twice about whether it is worthwhile starting [a war] with us," he said.[13]

Ariel Sharon, however, was among those who did not believe in the value of mutual deterrence. During this period he declared, "Today there is a situation where small or medium-sized countries own weapons of mass destruction ... We are dealing with totalitarian regimes led by murderous men who could not care less about the fate of their people. You cannot deter those countries. Mutual deterrence does not exist at all."

Ehud Barak seemed to concur. Soon after the first Gulf War, Barak noted that "the very intention" of countries such as Algeria, Iran, Libya and Pakistan to develop a nuclear capability, "is very troubling, even if these countries still face long years before their intention materializes ... This further stresses the importance of nipping these efforts in the bud, as early as in their Iraqi version. These efforts must absolutely be cut off."[14]

Israel's efforts since the first Gulf War to push the United States into a more confrontational stance towards Iran have been gratified, most clearly in recent actions led by the administration of George W. Bush. Iran has been marked in both capitals as the most potent regional threat to "stability" and its brand of Islamic "fundamentalism" has been identified as the foremost regional enemy not only of the West but also of every Arab regime from Saudi Arabia to Egypt.

However, the road to the current US policy has not been a straight one. During the government of Benjamin Netanyahu, Sneh called President Clinton a modern-day Neville Chamberlain, after Washington waived sanctions on companies investing in Iran's South Pars gas field. Sneh decried the move as "the appeasement syndrome that existed at the time of Hitler."[15]

This campaign has traditionally been burdened by a lack of accurate intelligence and consistent motivation. As a consequence, for more than a decade there have been claims that Iran has been on the threshold of obtaining an operational nuclear weapons option.

The strategic significance to Israel of an Iranian nuclear weapons capability has implications not merely for these two countries but for the region as a whole. Israel's vast nuclear arsenal, constructed outside the regime of the nuclear Non-Proliferation Treaty, is multi-dimensional and fully integrated into strategic and war fighting plans. On at least one occasion, serious consideration was given to employing its capability on the field of battle.[16]

Israeli concerns about Iran's nuclear program established an attractive rationale for an Israeli effort during this period to develop a credible "second strike" nuclear capability to counter the anticipated development of nuclear weapons by Iran and perhaps Iraq. In the lexicon of nuclear strategy, a second strike capability exists when some portion of the nuclear forces of a country are able to survive a non-conventional attack.

David Ivri, a longtime official at the center of Israel's defense policy, explained that Israel's nuclear arsenal must survive any attack by a nuclear power employing ground to ground missiles. "The development of deterrent power which will cause the other side to refrain from a surprise attack. Whoever plans a surprise attack using ground to ground missiles has to understand that if he doesn't succeed in destroying the entire offensive capability of the other side, he can expect a massive blow from which he will not recover."[17]

In the Iranian context, Ivri's message was that the development of a nuclear capability, and the missiles to deliver it, would not enhance Iranian capabilities vis-à-vis Israel. He was informing Damascus that its Iranian connection was a strategic dead end as well as a political and economic calamity, while indicating to Washington that Israel, far from planning to shed its nuclear weapons, was actively implementing a strategic view of the region requiring the development of new, more advanced capabilities.

The anticipated development of a nuclear arsenal in Iran, however, as well as Israel's experience during the Gulf War, created a new set of considerations. An Israeli decision to create a dedicated second strike capability presumes that Israel's nuclear monopoly will soon be ended and replaced with a bipolar or multipolar nuclear Middle East in which Israel is threatened by nuclear-armed ballistic missiles. In this environment, nuclear deterrence – and the creation of a second strike capability should deterrence fail – would become a favored Israeli option.

This transformation of the strategic environment was foreshadowed during the Gulf War, when Israel and Iraq were forced to calculate the costs of employing non-conventional weapons. Yet while Israel could, and did threaten Iraq with a nuclear response to a non-conventional attack, Iraq lacked a nuclear – but not a chemical – weapons option.

Israel considers its nuclear weapons to be a preeminent factor in its regional power. For example, Foreign Minister Ehud Barak considered, "the perception of Israel in the Arabs' consciousness as having nuclear capability" to be its foremost strategic asset. However, as Ivri's recommendation suggests, simply having the bomb is no longer a sufficient deterrent to attack in an era when Israel's nuclear monopoly is under threat.

Even as policymakers and pundits debate the plausibility and effectiveness of a pre-emptive strike against Iran's nuclear infrastructure, Israel's creation of a second strike capability represents recognition of the failure of its preferred policy of preemption, most visibly employed by the destruction of Iraq's Osiraq nuclear reactor in 1981. It also represents an acknowledgment that the United States has failed to keep the nuclear genie in its bottle.[18]

The new environment created by the Iranian efforts may hasten the day when Israel's nuclear capability is taken "out of the basement," where it has been kept not so very well hidden for more than a generation. This policy which once served Israeli interests so well has been overtaken by events. In order to protect the credibility of its nuclear arsenal in a nuclear environment, Israel may decide to adopt a policy of "open deterrence."

According to some analysts, this move to "open deterrence" may also imply a subtle change in Israel's nuclear doctrine. As long as the bomb was in the basement, Israeli leaders declared that Israel would not be the first to "introduce" nuclear weapons in the Middle East. The significance of Ivri's call for a second strike capability is that it suggests, at least to arms control experts, that Israel will not be the first to *use* nuclear weapons.

A strategy of preemption remained on the periphery of US considerations as long as there was a credible diplomatic-security framework for advancing US goals in the region. The Madrid process was understood as a means of isolating Iran and Iraq to form a regional consensus reaching from Morocco to Syria and the Arabian Gulf. The promise of Madrid exhausted itself with the failure of Israeli–Syrian diplomacy at Geneva in March 2000 and the implosion of Israeli–Palestinian rapprochement during the intifada that began in September of that year. A regional peace championed by the West was no longer seen as probable, nor the consolidation of an Arab–Israeli entente against "Islamic fundamentalism wielding nuclear weapons" that was to result from it. Bush and Sharon did not believe in an Arab–Israeli peace as the gateway to a more effective confrontation with Iraq and Iran. Indeed, the ongoing conflict between Israel and the Palestinians was viewed by many in Bush's Washington as a distraction, strategically unimportant and uninteresting. The failure of the Madrid process led to direct, unilateral efforts by the Bush administration to fix the region's stars in its favor, invading Baghdad and raising the concept of preemption of existing or nascent non-conventional capabilities to the level of strategic doctrine.

Israel's View of the Regional Strategic Environment

Its current abilities notwithstanding, Israel believes that its "strategic situation is better than it ever was ... compared to all the countries surrounding us."[19] It continues to be the region's sole nuclear power and has declared a strategic perimeter that ranges from Morocco in the west to Pakistan in the east and the Gulf in the south. US forces are engaged on Israel's eastern frontier, although the advantages to Israel of Saddam Hussein's removal are now being reconsidered by some Israeli policymakers in light of Iraq's chronic instability and the gains made in the country by forces allied with Iran. Israelis remain concerned about the regional implications of a US "defeat" in Iraq, which they consider to be a more probable outcome than a "victory." The "eastern front" – including Jordan and Syria – is quiescent, although some Israelis, particularly on the right-wing of the political spectrum, view instability in Iraq and Iran as validating Israel's continuing occupation of the Jordan Valley. Syria's military capabilities are atrophying and the peace with Egypt is holding.[20]

There is currently no regional power with the capacity to pose an existential threat to Israel. But Israeli policymakers consider an Iranian effort to gain a nuclear weapons capability a move which establishes Iran as a potential existential threat—particularly in light of the Iranian regime's vitriolic statements against Israel. However, Iran today poses no urgent challenge to Israel. The threats posed by the Palestinians and Lebanese (Hezbollah) are more immediate, if less serious.[21]

Strategic Doctrine

Israel's assessment of the implications of an Iranian interest in creating a nuclear capability must begin with an explanation of Israel's strategic doctrine; and since doctrine is first and foremost an expression of capabilities, let us begin by noting that Israel is a senior member of the nuclear club. It has a fully integrated arsenal of hundreds of nuclear weapons based upon the traditional triad of land-based, sea-based and airborne delivery systems, probably including a submarine-based second strike cruise missile capability.

Israel has never explicitly declared its nuclear capability; a policy of nuclear ambiguity evolved even before Israel attained an operational nuclear capability sometime in the 1960s, and was always married – after 1968 – to what was effectively an American guarantee of Israel's conventional superiority over any combination of Arab military capabilities. In the region, Israel views its nuclear monopoly primarily as a means to:

- restrain Arab war aims to something well short of the kind of military achievements that Israel would consider to pose an existential threat; and
- help to convince Arab antagonists that Israel's destruction is unobtainable and that as a consequence, diplomatic engagement on terms established by Israel is the only viable Arab option.

In the past, Israel has contemplated the use of nuclear weapons when its conventional capabilities were seen to be faltering—thus creating what was viewed as an existential threat to the nation. Such a consideration was made in the early days of the Syrian assault on the Golan Heights in October 1973, when a tank assault into Israel's heartland loomed.

It is important to note that this capability and doctrine has not prevented, or in some cases even deterred conventional attacks—notably in 1973 and more recently by Iraq. Arguably, Israel's non-conventional arsenal did deter chemical attacks by Saddam Hussein against Israel during the first Gulf War. There have recently been suggestions that Saddam Hussein considered Iraq's chemical weapons arsenal (real or imagined) to be a deterrent to an Israeli assault in the post-Gulf War period.

Over the years, Israel has maintained a serious and committed effort to preserve its nuclear monopoly in the region. Egyptian efforts during the 1960s were countered with a successful policy that included sabotage and assassination. A similar strategy was adopted towards Iraq in the 1970s and 1980s, culminating in the assault on Iraq's reactor at Osiraq. Success was far more ambiguous in this latter case. In the aftermath of Osiraq's destruction, Iraq reoriented its pursuit of a nuclear weapons capability, dispersing and hardening sites, and revising its strategy for attaining weapons-grade material. Only in the context of the unprecedented and intrusive post-Gulf War inspections regime was Iraq's nuclear file demonstrably closed.

Iranian Intentions and Capabilities

Is Iran truly determined to create a nuclear weapons option? In Israel's view this is a question that does not even need to be asked—the issue is not whether Iran is embarked upon such a strategy, but when, and at what stage of weapons development Iran is at.[22] For the last fifteen years Israelis and others have been warning that Iran is five years away from a nuclear capability. Israel's current assessment, at least the one noted in public, is that Iran is not yet at a point where it can create nuclear weapons. The operational use of weapons it does not yet possess is believed to be more than five years away. Israel, declared Acting Prime Minister Ehud Olmert on January 17, 2006, "cannot allow in any way or at any stage someone who has such hostile intentions against us to obtain weapons that could threaten our existence."

There are four options to be considered when evaluating the international response to Iran's efforts:
- *Destruction/Delay by Force*: Is it possible to destroy the Iranian program or to retard its development? Does Israel have the capability to undertake such an effort, alone or in combination

with other (US) forces? Is the solution worse than the disease? e.g., will the Iranian response to such a policy outweigh the advantages that its proponents foresee?
- *Destruction/Delay by Diplomacy*: This is the current and preferred method adopted by Israel and the international community. Some argue that this policy has already delayed the Iranian program.
- *Regime Change*: In some respects, Israel's concerns are less about Iran's capabilities than about the character of the regime—that is, who has their finger on the button. The current regime has failed to establish, and seemingly has no interest in establishing the kind of benign relationship that would encourage a change in Israel's view of the regime's intentions. Yet in this view, a non-Arab regime with a nuclear capability is not necessarily a bad thing—bearing in mind the "strategy of the periphery" that was the source of Israel's alliance with the Shah, whose interest in nuclear power was supported by the Nixon administration and who was allied with Israel in the development of nuclear-capable ballistic missiles. We would see a much more tolerant Israeli (and not only Israeli) view of Iran's efforts were they led by a different regime.
- *Accommodation*: Iran's attainment of a nuclear weapons capability would have significant repercussions on Israel's doctrine and could well influence the evolution of its nuclear arsenal. Iran's entry to the nuclear weapons club would end Israel's regional nuclear monopoly, although not its superiority; put pressure on Israel's policy of nuclear ambiguity; and undermine Israel's stated policy that it will not be the first to introduce nuclear weapons into the region. There would undoubtedly be regional implications as well. A more explicit Israeli declaration of its nuclear capability would put pressure on Israel to declare a doctrine regarding the use of its arsenal. Each of these developments would have regional

implications – in Egypt and perhaps the Gulf as well – and could well add to pressure on Arab states to seriously contemplate nuclear weapons initiatives of their own. Such a development would also affect the intricate relationship with Washington regarding Israel's nuclear capability. Upon assuming office, acting Prime Minister Olmert, like all previous Israeli leaders, noted his continuing commitment to maintaining Israel's deterrent capability, i.e., its nuclear option. Indeed, the crisis over Iran's suspected weapons program has prompted an unprecedented public commitment by the US President to come to Israel's defense in the event of a stand off with Iran.[23] The Bush remarks suggest that the US nuclear umbrella is now more explicitly extended over Israel—at least as far as Iran is concerned. This policy may protect Israel, but may also inhibit its freedom of action. Indeed, the loss of Israel's autonomous ability to undertake strategic military actions against enemies – from the decision not to enter the first Gulf War to the more recent desire to see the United States lead efforts to prevent Iran from going nuclear – defines the current era.

In what other ways will an Iranian nuclear weapons capability affect Israeli war-fighting and conventional policies? Will there be a nuclear dimension to Israeli attacks against Hezbollah in Lebanon? In other words, will Israel refrain from actions against Hezbollah because of the deterrent effect of an Iranian nuclear capability? Will such an Iranian capability exercise a deterrent effect more convincing than those aspects of deterrence that already exist and have proven effective in the years since Israel's withdrawal from the south?

Indeed, Israel's antagonistic relationship with Tehran, and more specifically Iranian desires to establish a deterrent to Israeli aggression against it, has long been viewed as an important factor in Iran's support for enhancing Hezbollah's missile arsenal. "This [surface to air missile] development is very dangerous and constitutes a direct and new threat on Israel," explained Dr. Shimon Shapira, a brigadier general who has been military adjutant or intelligence adviser to three Israeli Prime Ministers – Yitzhak Rabin, Shimon Peres and Benjamin Netanyahu – and author of *Hizbullah: Between Iran and Lebanon*. "Unfortunately the Iranian missiles arrived in Lebanon, among other things, because of various people in Israel who in recent years threatened to attack targets deep inside Iran," he claims.[24]

Concerns about Iran's efforts in the nuclear domain have led to a growing international campaign aimed at short-circuiting not only illicit Iranian weapons activities but also activities not prohibited under the NPT. Israel has supported these efforts – indeed it has championed them – but if the international community mobilizes against Iran, there is also, in Israeli eyes, the "danger" that the intrusive gaze of the international community will focus on its own nuclear capabilities as part of a revived effort to establish a regional diplomatic framework for nuclear disarmament. There was a hint of this in the February 4, 2006 IAEA statement calling for a Nuclear-Free Zone (NFZ) in the region.[25] Yet Israel is not prepared to enter into a framework that would establish any sense of symmetry between its nuclear weapons and those of its regional antagonists.

The potential for Israeli–Iranian rapprochement cannot be dismissed as impossible, although admittedly current affairs are more polarized than at almost any time since the Iranian revolution. It is worth recalling that Israel's May 2000 withdrawal from southern Lebanon opened a short-lived Israeli debate on Iran. In April that year, Minister of Justice Yossi Beilin became the first top level Israeli official to call for a rethinking of Israel's antipathy toward the Iranian revolution. In an April 4, 2000 speech at Haifa University, Beilin declared that "the Iran of President Khatemi and Iran after the elections is a country with far more nuances and far more complexity than we have become accustomed to see. Due to the positive changes evidenced in Iran, there is a need to change our approach towards them. An opportunity for a new opening is at hand." Beilin called for a "reassessment" of Israeli policy, particularly given the fact that both Iran and Israel "had and have common regional interests."

"We should examine our attitude towards Iran," he continued, "we had good and special relations with Iran until 1979, and we still have mutual regional interests. The problems began under Khomeini, who used Israel as glue to try and create social cohesion within his own country. I do not want to look at Iran with the same glasses that I was using 20 years ago."

Israel may be determined to see Iran denied a nuclear weapons option, but it is not prepared to pay for it by surrendering its own arsenal. Israel's nuclear option has always been a source of power in its relationship with Washington—from the Johnson administration's decision to assure Israel's conventional superiority after the June 1967 war, to Secretary of State Henry Kissinger's consideration of Israel's military re-supply during the October 1993 war and the extraordinary attention paid by the administration of George Bush to destroying Iraqi Scud sites during the first Gulf War.

Indeed, one factor compelling US policymakers to move against Iraq's invasion of Kuwait may have been their fear that Israel, unable to countenance such a successful assertion of Iraqi power, would wage a war against Iraq in which non-conventional weapons might be employed.[26]

Nuclear weapons are at the heart of Israel's strategic doctrine. Israeli proponents of nuclear deterrence, analyzing Iraqi behavior during the first Gulf War, argue that nuclear deterrence did in fact work and must be maintained.

There is little doubt that Israel, under the leadership of Ehud Olmert, supports the declaration of former Prime Minster Yitzhak Shamir, who in an address before Israel's parliament on October 7, 1991, warned Washington that Israel would "insist on maintaining reasonable security margins that will allow a secure life in the future." In the subtle discourse in which Israel's official nuclear debate is conducted, "reasonable security margins" translates into a vibrant nuclear option.

Conclusion

What options exist for Israel to address Iran's presumed, emerging nuclear weapons capability? In assessing options, it may be useful to note the history of Israeli actions against Egypt and Iraq. The former were successful, contributing to the termination of the Egyptian program. In the latter, Israel's attempt to physically destroy the Iraqi program did delay it, but it also made subsequent Iraqi efforts far less transparent. Iran has taken this lesson to heart, complicating if not making impossible any attempt to preempt or perhaps even significantly delay the Iranian program by force.

Israel has an interest in seeing that the NPT regime works—it has no interest in creating an environment where other NPT signatories in the region feel compelled to consider creating a nuclear option. An Iranian ability to create a nuclear weapons program while under the NPT would undermine the assumptions of a generation of arms control among its signatories in the region.

Israel also has a long-standing interest in not being the prime actor in efforts to retard – by force or diplomacy – or destroy the Iranian nuclear program. Israel's efforts to call attention to the program in order to generate concern by stronger powers appear to have succeeded. However, it has not yet succeeded in compelling Iran to adopt greater transparency and cease its controversial programs. Israel has established the parameters of the international diplomatic effort. It has suggested the instruments available to the principal powers – the EU-3, Moscow and Washington – and lobbied for their adoption. However, it has let others take the lead. "We are giving priority at this stage to diplomatic action," explained Defense Minister Shaul Mofaz on January 21, 2006, "but in any case we cannot tolerate a nuclear option for Iran, and we must prepare ourselves."[27]

5

Bombing Iran: Is it Avoidable?

Sverre Lodgaard

Since the events of September 11, 2001, it has been US policy "to stop rogue states and their terrorist clients before they are able to threaten or use weapons of mass destruction against the United States and our allies and friends." This policy emphasizes preventive action, or proactive counter-proliferation. To defend itself, America may have to launch "periodic" wars, and if necessary to do this "alone." This is often referred to as the Bush doctrine, elaborated upon in – among other places – the National Security Strategy of September 2002.[1]

Regime change is a primary objective of this strategy—in Iraq, Iran and North Korea, but also in Syria and other so-called "rogue states." The objects of elimination are hostile regimes, rather than weapons of mass destruction or any particularly objectionable behavior. The US attitude towards the possession of nuclear weapons depends on the states which possess them—for the time being, no objection is voiced against Israeli nuclear weapons, while proliferation in other Middle East states is deemed unacceptable.

The policy of regime change can also be understood in terms of US empire-building. During George W. Bush's presidency, the focus has been on the Middle East. Most Arab regimes are already oriented towards the Western world and the United States in particular. After the overthrow of Saddam Hussein and the occupation of Iraq, only Iran and Syria remain recalcitrant. The leadership in Tehran and the Assad family in Damascus are under strong pressure, but workable strategies to substitute US-friendly governments for them have proven hard to devise.

This approach to empire-building resembles the classic Leninist description. It is a "bridgehead" approach, installing leaderships that will work with the imperial power while keeping the grassroots at bay. In Lenin's formulation, the *center of the periphery* (i.e., the governments of the Middle East) develops stronger common interests with the *center of the center* (Washington) than with the *periphery of the periphery* (their own people).

When President Bush declared victory in Iraq in the spring of 2003, the overthrow of Bashar Al-Assad and Ali Khamenei may have seemed within reach—the threat or use of force – cf. "periodic" action – would make it happen, or it would come about in less dramatic ways. This turned out to be mere hubris. The ongoing war in Iraq soon undermined the vigor with which the United States could pursue regime change in neighboring countries. Therefore, the United States had to go about its objectives in a softer manner. For a couple of years, the US has been in "diplomatic mode." However, there is nothing to suggest that the administration has relinquished its goals, or that the basic tenets of its strategy have changed. On the contrary, the updated March 2006 version of the National Security Strategy reconfirms the same basic guidelines and clarifies their operative meaning in the current circumstances.

The update has a preface signed by the President, the first words of which are: "America is at war." It names seven tyrannical regimes: North Korea, Iran, Syria, Cuba, Belarus, Burma and Zimbabwe. Two of them are singled out because they continue to harbor terrorists and sponsor terrorist activity abroad – Syria and

Iran – while one of them is singled out for trying to acquire nuclear weapons—Iran. However, according to the document the concerns about Iran are much broader: "it threatens Israel, seeks to thwart Middle East peace, disrupts democracy in Iraq, and denies the aspirations of its people for freedom." The conclusion is that Iran represents the single greatest threat to the United States.[2] The document therefore enhances the focus on Iran as the next possible target of US warfare.

However, bombing is not a recipe for regime change. On the contrary, when nations are under threat, people usually mobilize in support of their leaders. Domestic conflicts are set aside in defense of a higher cause. So why does the United States deem it vitally important to take military action – if necessary – to stop Iran from acquiring nuclear weapons?

This stance is partly due to the history of the relationship between the United States and Iran, which is highly politicized and deeply adversarial. There is a bipartisan animosity towards the ayatollahs and Iran is the most difficult country for the United States to engage diplomatically. Insert Iran's nuclear program into this adversarial relationship and it becomes even more confrontational—especially since the program happened to surface in a fundamentally new international context driven by 9/11 and a much more assertive US policy. Nobody in Iran, or anywhere else, could have envisaged this—it was a historical coincidence of sorts. Today, the US conviction that Iran is secretly planning to develop nuclear weapons, and Iran's categorical denial of this, have plunged relations between them to their lowest state since the hostage crisis in 1979. No US government will accept Iran becoming a nuclear-armed state. Before it comes to that, military action will be taken.

Another aspect of the US stance relates to the physical control of oil supplies. One third of the world's oil supply flows through the Strait of Hormuz, and to keep it flowing has been the bedrock US foreign policy for more than 50 years. Mohammad Mossadegh was overthrown partly due to an unseemly affinity to the Iranian communist party (the Tudeh party) and partly because of his plans to nationalize the Iranian oil industry. The Shah's unswerving commitment to the free flow of Iranian oil became a central pillar of the Nixon doctrine. Likewise, in his final State of the Union address, President Carter declared that, "Any attempt by any outside force to gain control of the ... [Arabian] Gulf region would be regarded as an assault on the vital interests of the United States of America, and such an assault will be repelled by any means necessary, including military force."[3] The Reagan administration said the same, and began establishing military bases in Saudi Arabia. In 1990, when Saddam Hussein occupied Kuwait, Secretary of Defense Cheney stated: "We're there because the fact of the matter is that part of the world controls the world supply of oil, and whoever controls the supply of oil ... will have a stranglehold on the American economy."[4] If those considerations were not part of the reason for occupying Iraq, it would have been the first time in more than half a century that the uninterrupted flow of Gulf oil was not a central element of US foreign policy.

Today, geopolitics is first and foremost about energy supply and security. The United States has occupied Iraq and keeps a military presence in Afghanistan; has a number of bases in the Gulf region—including new ones in Iraq to replace those that were lost in Saudi Arabia; and deploys carrier groups in the vicinity of the Gulf. It holds the region in a tight military grip. However, a nuclear-armed Iran could question the credibility of that military dominance. Even if regime change is out of reach at present, it still makes sense to strike at Iran's nuclear program and limit Iran's ability to strike back.

A third aspect involves Israeli interests. Far from recognizing Israel, the Iranian President has said that he wants to wipe Israel off the map. To Israel, no threat is greater than a nuclear-armed Iran. Therefore, Israel tends to focus on the worst-case scenario when considering the Iranian nuclear program, claiming that Iran has an as yet uncovered separate, secret military program. Israel has prepared itself to take military action if nobody else (i.e., the United States) does. Israeli interests weigh heavily on US decision-making and Israeli leaders are no doubt leaning on the US to ensure that action is taken sooner rather than later.

The United States has always been ready to increase the pressure on Iran, and has been moving towards an escalation of the conflict, especially since August 2005, when Iran rejected the offer of the EU-3. However, it appears that it has pursued this course without having decided what to do in the event that non-military means fail to work. It has been argued that far from being a foregone conclusion, the United States has so many other problems on its hands that, if anything, it would try to avoid a military outcome. Others point out that politicians buy time when they do not know what to do—and this is what bombing can achieve. Bombing would set the Iranian program back and buy time for efforts to change the regime. In the long-term, bombing is likely to be counterproductive for both of these objectives—for example, it was *after* the bombing of Osiraq that Saddam Hussein mobilized his resources behind a comprehensive nuclear weapons program. However, politicians seldom plan beyond their election periods and policies are usually devised and conducted in much shorter time-frames.

There are also those who believe that the US leadership has already decided to use force. Military planning for this eventuality has been ongoing for some time; the White House has been consulting members of Congress on the matter;[5] and one can assume that Israel – silent for a while in the public domain – has been pursuing such an option behind the scenes. Likewise, in Iran the government may have decided to take what comes and ride it out. To a large extent, the present leadership belongs to the war generation of the 1980s, and has strong roots in the Revolutionary Guard. The hardliners obviously have the upper hand, while the so-called pragmatic conservatives, who care more than others about favorable international framework conditions for the Iranian economy, have lost much of their influence.

Yet others emphasize that the escalatory dynamics have grown so strong that they will be hard to break out of. They read the danger in terms of an inadvertent slide to war, reminiscent of the process leading to World War I, rather than a premeditated attack.

The Consequences of a Military Attack

The war against Iraq could be predicted with virtual certainty many months in advance. There were strong indications that the decision had been made, and in due course the preparations for occupation were unmistakable. The case of Iran is different. Preparations for bombing are not so easy to interpret. The aggressor will attempt to employ the art of surprise and therefore will do his best to mislead the adversary about the timing of an attack. While the probability that military force will be used seems high, it may still be avoided.

The advantages of halting the current escalation and achieving a political solution can only be fully assessed when compared to the costs of war. These costs cannot be predicted with any precision, but they are potentially huge, not only for the warring parties, but for the entire Middle East region, as well as for energy prices and economic development worldwide. Compared to the case of Iraq – where realistic assessments of the long-term consequences of war were absent – much attention has been drawn to the political implications of the use of force against Iran. The lessons from Iraq are sobering. If war is indeed the endgame of the current escalation, which war scenarios are likely and what will the political and military consequences be?

Likely Targets of a Military Attack

If it comes to war, will the attack be carried out by Israel and/or the United States? Clearly, different capabilities translate into different war scenarios.

Israel has long-range F-15s and F-16s that can be used to bomb nuclear installations in Iran. It can also use cruise missiles on board submarines. Furthermore, it has exercised mock attacks on a Natanz-like target erected in the Negev desert, so incursions into Iran from Northern Iraq to destroy dual-use enrichment technology may be on the cards. Israel has a foothold in Kurdish-controlled territory in the north from which it can operate.

However, what Israel can do is of a relatively small scale in comparison to US capabilities. The US can use aircraft from CONUS, the UK, Diego Garcia, its bases in the Middle East and from aircraft carriers. Normally there are one or two carrier groups in the area at any given time, each with some 75 aircraft on board. A variety of cruise missiles are available and long-range ballistic missiles with conventional warheads may also be used. The Nuclear Posture Review of 2001/2002 announced a new triad of strategic forces consisting of offensive strike systems, defenses and a revitalized defense infrastructure, while offensive systems have been reconfigured to include both conventional and nuclear-armed ballistic missiles.[6]

For Israel, the targets are nuclear installations, while for the United States, other important objectives are the destruction of military infrastructure – that of the Revolutionary Guard in particular – and actions to control ship traffic through the Strait of Hormuz. To protect oil transports and near-by oil installations, the Iranian side of the Strait may be occupied. The magnitude of such a

US attack may not be much smaller than the opening salvos of the war against Iraq.

The timing of any attack will be chosen in order to maximize the element of surprise—preparations for air attacks are elusive enough to give the attacker a certain advantage in this respect. For Israel, surprise is important to reduce the effectiveness of Iranian countermeasures, which are more easily dealt with by the United States. Both would like to hit scientists, engineers and technicians working in the facilities—it is only prudent to assume that Iran will try to reconstitute its nuclear program as best it can, so decimation of its qualified workforce is of the essence.

Iran is building tunnels and cavities in hard rock to prevent valuable assets from being destroyed. This is vital for the rapid reconstruction and resumption of nuclear plants and activities. These shelters may be hard to destroy by conventional means, so the use of nuclear weapons may be considered. The Bush administration has tried to erase the distinction between conventional and nuclear arms, and its nuclear doctrine says that the United States may use nuclear weapons even in situations where the adversaries only possess conventional weapons. It also maintains a keen interest in new ground penetration warheads, although Congress has been unwilling to fund new "bunker buster" programs. In Iran, there are targets in hard rock that may be invulnerable to conventional means. Therefore, it should come as no surprise if the Bush administration actually does what it has been writing and talking about. On the other hand, the use of nuclear weapons was considered several times during the Cold War, but the conclusion was always to abstain. If the idea is dropped once again, this should come as no surprise either. The point is that the question is considered worth posing.[7]

Both Israel and the United States have been planning and exercising for a military attack. When political leaders emphasize that all options are on the table, this is implied. If not, it would merely be empty talk. This time, preparations have gone further, beyond anything that could be thought of as "business as usual." However, even if thorough planning has been done and superior forces are available, unexpected difficulties may arise. Complex operations never unfold in quite the way they are planned to. Nevertheless, in comparison to the long-term political and military *consequences* of an attack – for the Arabian Gulf and the wider Middle East – a military campaign is predictable. Bombing is a time-limited activity which leaves no ground forces in a quagmire.

Consequences of a Military Attack

A military attack may be staged in combination with covert operations to change the regime in Tehran.[8] US covert operations had a notable success in 1953, when Mossadegh was overthrown. Another success seems improbable, however. It is more likely that the Islamic Republic will survive and gain even broader popular support than before. The nuclear program is backed by all political groups in Iran. While there is much in-fighting on many other issues and different views on the specifics of nuclear policy, the right to develop a comprehensive program for the utilization of nuclear energy has near universal support. Bombing will narrow the scope for legitimate dissent and close the ranks behind the leadership.

Precision bombing can reduce civilian losses. Nevertheless, such losses will be large, for many targets are located in or near cities. In the aftermath, the nuclear inspectors will no longer have access to facilities and the nuclear program will be restarted. If Iran was not motivated to build bombs before, it will be determined to do so afterwards. Signs of renewed activity may trigger new bombing raids similar to those in the no-fly zones over Northern and Southern Iraq after the Gulf war of 1991. This is a recipe for continued high tension in the region, blocking initiatives for confidence-building and arms control.

In recent months, conflicts in the Middle East have intensified. Hamas won the Palestinian elections; developments in Lebanon and Syria are tense and unpredictable; the violence in Iraq continues unabated and may even escalate; and at the grassroots level, large segments of the Arab world are becoming radicalized. The bombing of Iran will therefore interact with other regional conflicts in ways that are hard to predict—thus this is a particularly hazardous moment at which to start another war.

Obviously, Iran can stimulate unrest in Iraq. There are close ties between the Shia populations of the two countries. Indeed, some Iraqi leaders are even citizens of Iran. Through its Shiite partners in Afghanistan, Iran can provide arms for warlords and undermine efforts to stabilize the country. It will not be able to close the Straight of Hormuz, but individual acts of sabotage against ship traffic may succeed. Oil facilities in Saudi Arabia, Kuwait, Qatar and the UAE may also be sabotaged. In addition, Israel is likely to be targeted, directly from Iran and/or through Hezbollah. The Revolutionary Guard is the main Iranian instrument both in Iraq and in its support of the Palestinians against Israel. Bombing will engage 1.2 billion Muslims – many of them emotionally – and international terrorist networks will be emboldened to act in revenge.

Iran's ability to escalate these and other conflicts is often used by those who oppose war in the Middle East, Europe and elsewhere to dissuade the US and Israel from launching a military attack. However, the decision on whether or not to use force will be made in the United States (and Israel), i.e., in political landscapes that are quite different. US opinion polls show support for bombing in the range of 40–60 percent, while in Congress the animosity towards the Iranian leadership is intense and bipartisan. There is broad support for the stance that Iran should never be allowed to become a nuclear-armed state and US leaders are turning some of the war opponents' arguments around by emphasizing, for instance, that much of the unrest in Iraq is caused by the Revolutionary Guard, so its capabilities had better be cut. However, the political market is nervous in view of the upcoming Congressional elections and the impact of another war on petroleum prices, and poll results obviously depend on how questions are posed. Israel realizes that Iran can hit back, but has its priorities clear: anything Iran can do in response is less threatening than an Iran armed with nuclear weapons.

Disagreements and Missed Opportunities

The major powers read the Iranian challenge differently; they pursue different objectives with regard to Iran; and they therefore differ in their approaches to the challenge. These differences can be analyzed along a number of lines. In the most rudimentary of terms, there is a dividing line between the EU-3/EU and the United States on the one hand, and Russia and China, supported by non-aligned states on the other.

Disagreements Among Major Powers

Disagreements between the major powers account for much of the failure to achieve a peaceful solution to the Iran issue. Iranian leaders can live with antagonistic US policies – and possibly even better *with* US sanctions than without – but had the outside world been united and firm in demanding a halt to all fuel cycle activities, it might have succeeded. Iran is one of the most open societies in the Middle East. Comprehensive isolation – people seeing that the world was turning against them – could have triggered popular protests against current policies. The Iranian government therefore did its best to avoid being judged by the world organization. In order to prevent this, it sought alliances with the Europeans, Russia and China against the United States and Israel. Reporting a country to the United Nations Security Council singles it out as an unreliable actor on the world stage: for the heirs of the proud Persian civilization, this is humiliating.

However, by the time the case of Iran was brought before the Security Council, which asked it to stop all enrichment activities; provide extended access for the IAEA inspectors; and accept all the other demands of the IAEA Governing Board, it was clear that if Iran stayed put, it would be hard for the Permanent Five (P-5) to agree on compulsory Chapter VII action. They would not agree on comprehensive economic sanctions and certainly not on the use of force. This made it easier for Tehran to do what it said it would do—continue its enrichment works and withdraw from the Additional Protocol to the safeguards agreement.

The EU-3

In hindsight, a number of opportunities for the peaceful resolution of the conflict have been missed. Some of them were simply not pursued, while others may have been better explored.

The main attempt to reach a diplomatic settlement was made by the EU-3/EU – France, Germany, the United Kingdom and the High Representative of the European Union – which sought a political solution within the framework of the non-proliferation regime. However, Iran's response to the Framework for a Long-Term Agreement offered by the EU-3/EU on August 5, 2005 was stiff: "the proposal is extremely long on demands ... absurdly short on offers to Iran ... [and] amounts to an insult on the Iranian nation."[9]

At the core of the offer were assurances of fuel supply for Iranian power reactors in return for a halt to all fuel cycle activities in Iran. The proposal elaborated on how fuel supplies could be assured in practice, expressing support for cooperation between Russia and Iran and committing the EU-3 to assist in the establishment of a buffer store of fuel, sufficient to maintain supplies for a period of five years. While international supply arrangements can never be as reliable as domestic sources of supply, and the buffer store would be located outside Iran, the credibility of these assurances was high. They were made by a group of states and communicated to all interested parties through an international organization (the IAEA), and so could not be withdrawn suddenly by any single government. It was suggested that the IAEA "might be invited to monitor the operation of the mechanism and certify its operation on objective principles."[10]

The Framework recognized Iran's right to develop a nuclear power program in order to reduce its dependence on oil and gas and to choose the most appropriate mix of energy sources. However, it stopped short of offering Iran light water reactors. While Article IV of the NPT commits supplier states to facilitate access to technology for Non-Nuclear Weapons States (NNWS) parties, the Framework only promised "not to impede participation in open competitive tendering."[11]

According to the Framework, in addition to stopping all fuel cycle activities, reconfirming its NPT obligations and ratifying the Additional Protocol on safeguards, Iran should undertake to cooperate proactively with the IAEA to solve all outstanding issues, "including by allowing IAEA inspectors to visit any site or interview any person they deem relevant to their monitoring of nuclear activity in Iran." In response, Iran noted that such inspections would go beyond the Additional Protocol, and considered this demand an intimidating infringement on its sovereignty.[12]

The Paris guidelines of November 2004, on which the negotiations were based, said that the agreement would provide "firm commitments on security issues." However, the offer did not; it merely referred to the UN Charter and reaffirmed the security assurances that France and the United Kingdom had given together with the other veto powers, summarized in Security Council Resolution 984 of 1995. The EU-3 reaffirmed their commitment to work for a zone free of weapons of mass destruction in the Middle East, but without introducing any new element that would take that proposition forward. As part of an overall agreement, the EU-3 welcomed an expanded dialogue on regional security issues. However, none of this addressed Iran's security concerns in a "firm" manner.

Neither was it easy for the EU-3 to do so, for the main threat to Iranian security comes from the United States, which took an ambivalent view of the negotiations. When working on the Paris agreement, the Europeans deemed it important that the United States should be comfortable with its provisions. The precise terms of the suspension were important in that respect. In the negotiations that followed, they kept the United States well informed. However, far from considering any security assurances for Iran, US pressure on the ayatollahs continued. Commenting on the Iranian rejection of the EU-3 offer, President Bush ended up referring to the use of force as a means of last resort.[13]

In the early stages of the negotiations, the United States made two gestures: it would no longer object to Iranian negotiations for WTO membership – which had been on the Iranian demand list to the Europeans – and it was willing to provide spare parts for Iranian civilian aircraft. There is a long road, however, from the beginning of WTO negotiations to a successful conclusion and a great many ways in which such talks might derail. Far from meeting Iran's main concerns, Tehran scoffed at Washington's gestures. To the Iranians, it confirmed the unyielding hostility of the United States: if this was all it was willing to do, it meant that the United States was not interested in facilitating a deal.

In practice, the US offer was addressed to the Europeans as much as to the Iranians. It conveyed a semblance of support for European diplomatic endeavors while remaining at a distance. Since the Iranian fuel cycle activities would be suspended for as long as the negotiations lasted – and the suspension was defined in accurate and comprehensive terms – diplomacy bought valuable time for the Americans. As long as the Europeans stayed committed to halting all fuel cycle work in Iran – which they did and still do – the talks could do no harm and could only be helpful. However, the United States never gave them much of a chance beyond being a holding maneuver.

The Framework for a Long-Term Agreement was too little too late. At an earlier stage of the negotiations, it could have made sense as an input for further consideration, but five days after the August 1 deadline it had no chance of winning Iranian acceptance. The European negotiators could hardly have failed to foresee that, although they may have been surprised by the strong-worded, categorical nature of the rejection.

Would the Iranian reaction have been different had the Europeans offered power reactors and been more forthcoming on security assurances? If the EU-3 had been quicker in tabling their proposals – Iran was annoyed by the slow pace of the talks – would the negotiations have continued and options for a political solution been better explored? The answer to this remains unknown, but in hindsight the restrained European attitude may seem regrettable, because it left the feeling that there was more to discuss.

Other Diplomatic Initiatives

August 2005 was a turning point. Iran briskly rejected the EU-3 offer and in a comment President Bush ended with a reference to the use of force; to which Chancellor Schröder said that under no circumstance would Germany participate in an armed attack. Clearly, the conflict was set on a path of escalation.[14]

In 2005, an IAEA expert group reported on the problems and possibilities of multinational fuel cycle arrangements. When releasing the report, the chairman of the group said he believed that arrangements of this kind could provide assurances that enriched uranium would not be diverted to a clandestine nuclear weapons program.[15] In addition, a number of technical fixes could significantly enhance their proliferation resistance.[16]

Two of the options identified by the group are based on the notion of shared ownership or control, promoting multinational arrangements with the participation of other nuclear and non-nuclear states as confidence-building measures. Upon taking office, President Ahmadinejad entertained the same idea, proposing that public and private companies might be involved in the sensitive parts of Iran's nuclear program.[17] Internationalization of the nuclear fuel cycle was discussed 25 years ago – Europe already had two multilateral uranium enrichment plants, Urenco and Eurodif – but had not made any headway in regions of real proliferation concern. Today, the need for such arrangements is stronger than ever before because nuclear power is on the rise and the international consensus that fostered the NPT is about to break up. Iran represents an opportunity to explore multilateral options where they are most needed.

Had IAEA inspections been allowed to run their course and the Agency come to a satisfactory conclusion, a multinational fuel cycle arrangement could have been the next phase in a peaceful resolution of the Iranian nuclear issue. At the very least, the pursuit of multilateral options would have put the Iranians to the test—if they were serious about limiting the fuel cycle to civilian uses and enriching their uranium in a multilateral setting, they should have accepted the voluntary transparency measures that the Agency asked for in order to "come clean." A multilateral arrangement of this kind presupposes that the partners are confident that no other enrichment works are taking place under host country auspices.

However, the IAEA's efforts to reconstruct Iran's nuclear history and clarify whether there remained any undeclared nuclear activity or facilities in the country were interrupted. Like Iraq in 2003, where the inspection process was overtaken by the urge to go to war, the Agency's technical and diplomatic efforts to complete the mapping of Iran's nuclear program were overwhelmed by the politics of the matter. Before the Agency was ready to conclude, the issue was reported to the Security Council, whereupon Iran went back on the Additional Protocol, making the IAEA unable to search for undeclared items. A qualified conclusion saying that there are no undeclared facilities and activities in Iran, and that fissile materials have not been diverted for weapons purposes, would have made it easier to deal with the remaining issues. However, that possibility was discarded.

For a few years, Iran accepted more comprehensive inspections than any other country in the world. It gave the inspectors access beyond the requirements of the Additional Protocol, albeit reluctantly.[18] The United States gave some leads for the Agency to follow up on, and so kept it busy in a way reminiscent of UNMOVIC in Iraq, but without resulting in the discovery of anything hidden. Iran was left with the impression that however much transparency it offered, it would never be enough.

At the turn of 2005/2006, talks were ongoing between Russia and Iran. Russia invited Iran to cooperate on uranium enrichment on Russian soil. Iran, however, wanted to combine industrial-scale enrichment in Russia with pilot-scale enrichment in Iran, often referred to as research and development activities. At the core of these talks were the *place* and *time* of enrichment. Industrial-scale enrichment would take place in Russia, but to what extent would

Iran be involved? What would be the role of Iranian scientists? Would pilot-scale enrichment in Iran, involving only a small number of centrifuges so that the activity would be harmless from a weapons point of view, be agreeable? For how long or under what circumstances would enrichment have to be performed in Russia, and when or under what conditions could industrial-scale enrichment take place in Iran?

Separative work units (SWUs) – measures of enrichment capacity – are a continuum: at the low end, the amounts are militarily insignificant. For instance, if Iran were allowed to run UF_6 through the cascade of 164 centrifuges that had been set up at Natanz – and was limited to that – the amount would be militarily insignificant. The *degree* of enrichment could be limited to, for example, no more that five percent – i.e. what is needed for reactor fuel – and the production would be under IAEA surveillance. For the Western powers, however, *no* enrichment on Iranian soil was acceptable. They probably had in mind a moratorium on enrichment and reprocessing in Iran which would last for up to twenty years or so. Russia, too, stayed opposed to enrichment on Iranian soil but was not specific about the time frame of a moratorium.

The length of such an agreement was not necessarily a matter of chronological time, i.e., a certain number of years. It could last until IAEA inspections came to a positive conclusion. This would not have been a quick fix. Even under normal circumstances, determining whether everything has been declared on the basis of the rights and obligations provided by the Additional Protocol is a time-consuming process. It took five years in the case of Japan and three and a half years in the case of Canada. Also, the conclusions are not without reservations—the Additional Protocol is a big step ahead for international safeguards, but it does not lead to a foolproof, clean bill of health.[19]

Prior to the meeting of the IAEA Governing Board on March 6, 2006, Russia indicated that Iran might nevertheless be allowed to conduct small-scale enrichment works—small enough to be militarily insignificant. The Director General of the IAEA had similar thoughts, realizing that in order to strike an agreement with Iran, some face-saving measure would be necessary. Germany may have entertained the idea, but France and the UK were negative, and the United States rejected it out of hand. After a while, Russia said that it would not introduce such a proposal, and the EU-3 maintained its position not to accept any degree of enrichment in Iran. It was at this moment that Iran proceeded to circulate uranium gas in the centrifuges that had been installed at Natanz, while limiting the IAEA's presence to the provisions of the standard safeguards agreement (INFCIRC/153). Thus, another chance to reach a political agreement had been missed.

In summary, the offer by the EU-3 was too little, too late; the multilateral fuel cycle option, proposed by Iran itself, was not pursued; the US support for the inspections process was less than genuine—wanting to enhance pressures on Iran, a positive conclusion that nothing had been hidden was not in its interest, however qualified; the talks about joint enrichment on Russian soil had failed; and ideas about a face-saving compromise for Iran in order to stop the matter from being transferred to the Security Council and avoid further escalation were quelled.

Throughout the process, the US objective of regime change made political solutions difficult. Non-proliferation and regime change are difficult to reconcile and policies of regime change make serious negotiations impossible. If one party makes it clear that its primary objective is to cut the throat of the other, the other has little incentive to negotiate. Furthermore, if one reads a weapons ambition into the Iranian program, there is a dynamic escalatory element in US–Iranian relations similar to the relationship between the United

States and North Korea: the United States threatens Iran; Iran pushes its nuclear program forward with a long-term view to keep outside powers from dictating to, and attacking it; and in turn the United States uses this to put additional pressure on the regime.

The Way Ahead

Is a political solution still possible? What might it involve? What are the parameters to consider? Having hidden the program for eighteen years and tried to mislead the IAEA through most of 2003, it would take a lot to re-establish confidence in Iran as a non-nuclear weapons state. This, however, has to be the starting point. Only when the IAEA has ascertained that there are no undeclared facilities or activities and trust has been re-established can Iran claim the full benefits of Article IV of the NPT.[20]

Ideally, a solution should be sought within the framework of the international non-proliferation regime and in a way that would strengthen the regime. Iran's acceptance of the Additional Protocol would help to establish it as the new verification norm. Acceptance of voluntary transparency measures could set a valuable precedent for clarification of the nuclear history of other states failing to live up to their obligations. If Iranian participation in a multinational fuel cycle center becomes part of the long-term solution, this would be the first time such a venture has been set up in a proliferation-prone region. Multinational centers may be the best way to bridge Article IV and Articles I/II of the NPT, making the peaceful utilization of nuclear energy more compatible with the non-proliferation objective.

For now, however, all of this has been overplayed. In essence, the matter has been moved from Vienna to New York, where it is subject to big power politics outside rather than inside the non-proliferation regime.

The Role of the Security Council

The IAEA Board of Governors decided to report the matter to the Security Council on February 4, 2006. The negotiations between six states (the P-5 plus Germany) that began a month later resulted in Council support for the IAEA resolution, urging Iran to re-establish a suspension of its enrichment and reprocessing activities; reconsider the construction of a heavy water reactor; ratify the Additional Protocol; and implement measures to increase transparency, as requested by the IAEA. This is exactly what the international body responsible for international peace and security should do, i.e., provide political support for the specialized organization verifying compliance with key provisions of the Non-Proliferation Treaty (NPT).

Iran was allowed thirty days to comply with the request (i.e., up to April 28), but there was never much doubt about the outcome—the request was unacceptable to Iran. By mid-April, it was running UF_6 through the cascade of 164 P-1 centrifuges installed at Natanz. Iran intends to have 3000 centrifuges in place there by the end of 2006 and President Ahmadinejad has announced that Iran is conducting research on second-generation P-2 centrifuges.

For a couple of years, questions have been raised concerning the whereabouts of P-2 technology – the technology came from the illicit Khan network in the mid-1990s – but Iran claims that it had its hands full with the application of the P-1 technology and therefore did not pursue P-2 centrifuges (except one minor activity).[21] Ahmadinejad's statement, however, reinforced suspicions that Iran has been doing more than it has declared.

The next step on the escalation ladder may be a Chapter VII resolution imposing sanctions. One possibility is smart or targeted sanctions; another option is an arms embargo. However, even if sanctions worked as intended, they would only do so over time. Therefore, it is questionable whether the United States would be particularly interested in them. Comprehensive iron-clad sanctions will be opposed by China and Russia, and agreement on military action is excluded. If the US and China can agree on a strategy, the Council might join by consensus, but it is very difficult to envisage what that strategy could be.

The Role of Russia and China

In an effort to ease the crisis, China also launched some diplomatic activity in search of a political solution, in parallel with Russian attempts. China has made major investments in Iran and has entered long-term oil and gas agreements with the country—China receives 14 per cent of its oil imports from Iran. The Chinese – worried about Security Council discussions on sanctions, further escalation of the conflict and unilateral military action – hoped to gain Iranian acceptance for a deal that would keep the matter in Vienna. The hope was also that the report that was to be submitted before the Board meeting of March 6 would convey a sense of progress in the clarification of remaining questions about the Iranian program. However, the efforts petered out.

Considering how other cases of non-proliferation concern have been dealt with, the fact that Iran has been reported to the Security Council does not necessarily mean that the Council will proceed to punish it. In recent years, the IAEA Board has reported Iraq, North Korea, Libya and Romania to the Security Council for possible censure because of their nuclear programs. North Korea has been reported twice, but in the face of Chinese opposition the Council has never voted to punish it. On one occasion, it did not even express concern.

In some important respects, China, Russia and Iran have common interests in this matter. China and Russia emphasize that Sino-Russian relations have reached "unparalleled heights," and that Iran is an important partner to both of them.[22] Along with energy supplies, arms transfers and investments, the triangle cultivates compatible foreign policies. On important issues like Taiwan and Chechnya they hold identical positions. China and Iran support Putin's war against the Chechen separatists and the recent promulgation of China's anti-secession law – stating Beijing's intolerance of Taiwanese independence in no uncertain terms – was heartily commended in both Moscow and Tehran. Another common denominator is their opposition to US unilateralism.

A joint statement from the Russia–China summit of October 2004 issued a strong rejection of the Bush administration's unilateral foreign policy. It noted that, "it is urgently needed to [resolve] international disputes under the chairing of the UN and resolve crises on the basis of universally recognized principles of international law. Any coercive action should only be taken with the approval of the UN Security Council and enforced under its supervision." The China–Iran–Russia triangle is a way to counter US global ambitions. Seen in this perspective, Iran is integral to the broader effort to thwart Washington's foreign policy goals. In view of this, China and Russia can be expected to oppose punitive action against Iran. Thus their last-ditch diplomatic efforts to avoid further escalation were no surprise.

Remaining Options

The possibility remains that in the face of stiff international reactions and threats of the use of force, Iran will back down and accept the demands made by the IAEA and the Security Council. However, to date there have been no signs that this will happen. On the contrary, by stating their unyielding position in so many words and on so many occasions, Iranian leaders have committed themselves very strongly to the nuclear fuel cycle program.

Another possibility is US–Iranian talks, not only about the nuclear issue, but about the situation in Iraq as well. Talks have been announced between the US Ambassador in Iraq and Iranian delegates. The United States has stressed that these talks will be limited to the problems in Iraq. However, to be of interest to Iran, and to give it a motive to be cooperative in handling the unrest in Iraq, the talks would have to be expanded to include the conflict over its nuclear program.

The chairman of the US Senate Foreign Relations Committee, Richard Lugar, has recommended direct talks between the United States and Iran. The suggestion may have been coordinated with the White House against the backdrop of possible bombing; the hope being that Iran would pick up the invitation and make concessions.[23] However, the administration categorically rejects anything of this sort.

As the Security Council is beginning to address the question of sanctions, US spokesmen say that the Iranian program has passed the point of no return: "Once you are able to operate, over a sustained period of time, 164 centrifuges in cascade and feed [UF_6] into that, you are well on your way to an industrial scale activity"— the figure 164 is claimed to be a key number.[24] By this standard, there is little time to wait for sanctions to have an effect—if they are at all achievable. To avoid the use of force, it seems that radical political reorientation of the kind indicated above would be necessary, in Tehran and/or in Washington.

A Zone Free of Weapons of Mass Destruction

The Iranian crisis involves global non-proliferation and counter-proliferation policies as well as geopolitics. It is firmly in the hands of Iran and the big powers and there isn't much that smaller states within and outside the region can do except to explore and support options such as those discussed above.

At the same time, there is a regional dimension to the Iranian problem, which involves Israel. It is an illusion to think that proliferation pressures in the Middle East will disappear without engaging Israel in terms of nuclear arms control. The proposal for a Middle East free of weapons of mass destruction recognizes this and so deserves particular attention.

The idea of a zone free of weapons of mass destruction in the Middle East was first proposed by Egypt and Iran in 1974. The 1991 ceasefire resolution on Iraq (Security Council resolution 687) explicitly linked the disarmament of Iraq to the establishment of such a zone. The 1995 review and extension conference of the NPT passed a resolution on the same theme. The resolution adopted by the IAEA Board of Governors on February 4, 2006 recognized that "a solution to the Iranian issue would contribute ... to realizing the objective of a Middle East free of weapons of mass destruction, including their means of delivery."

The approach would have to be an incremental one, moving step by step towards a comprehensive zone arrangement. However, nothing concrete has been done to set such a process in motion. Rather, the idea is conveniently referred to for purposes other than the realization of the proposal itself: in 1995, the Middle East resolution facilitated the decision to extend the NPT indefinitely, and the latest IAEA resolution inserted a reference to it in an effort to gain as much support as possible for reporting Iran to the Security Council.

Any breakthrough regarding the idea presupposes that all states in the region recognize each other. Today, this is not the case and negotiations about the provisions of a zone arrangement cannot start until Iran has accepted Israel's right to exist. Before it comes to that, however, coordinated unilateral measures could take the idea forward and in due course make negotiations easier to conduct.

In textbook logic, the Iranian crisis presents an opportunity to follow up on the idea in substantial terms. A first step could build on former President Bush's arms control proposal of 1991, in which he called for a freeze on reprocessing activities in Israel. Today, a freeze on the production of fissile materials in the Middle East would be aimed at Iran and Israel. Agreement to do this would amount to a regional cut-off in the production of fissile materials. In the case of Iran, such a freeze is at the core of the current dispute. If Israel could be convinced to institute a freeze in Dimona, it would be harder for Iran to refuse to do the same. More than that, it could be viewed as a diplomatic victory for Iran.

In the Middle East, a zone free of nuclear weapons – nuclear weapons being of greater military significance than chemical and biological weapons – would require particularly effective means of verification. Suspicions run so deep that there is a need for mutual and binding reassurances between participating states—meaning some form of mutual inspection above and beyond what the IAEA can offer. Such arrangements will not be easy to develop and agree upon. However, a ban on national enrichment and reprocessing facilities would be helpful in this respect, because it would create a "firegate" between civilian and military applications of nuclear energy, making verification tasks more manageable.

If George W. Bush were to follow up his father's proposal, it would send a powerful message to the Israelis. In essence, it would be the reverse of Yitzhak Rabin's strategic rationale for the Oslo agreements. Rabin said Israel should speed up the process of settling the scores with the Palestinians because in ten years' time, a graver security threat was likely to emerge from Iraq and Iran. Today, the threat from Iraq is gone, and the concerns about Iran may be put to rest through a binding cut-off agreement, subject to strict verification. It would be only reasonable to ask Israel to make a concession in return.

The harder part of that concession may not be the freeze, but its verification. Israel is averse to transparency and international inspections. Indeed, it maintains a policy of opacity in the nuclear sector, its declaratory policy having been the same all through its nuclear history—never to be the first to introduce nuclear weapons in the Middle East. The quest for identical obligations may therefore be hard to manage.

The United States represents another obstacle, both because of its support for Israel and because of the Bush administration's basic approach to proliferation problems. The President has said, many times over, that one must see to it that nuclear weapons are in the right hands. Therefore, for the time being, leaning on Israel seems out of the question.

Still, those who recognize the regional dimension of the problem should continue to press for textbook logic to also become good political logic, for unless something is done about the Israeli nuclear arsenal, non-proliferation will sooner or later become a doomed proposition in the Middle East.

CONTRIBUTORS

JOHN SIMPSON OBE is Director of the Mountbatten Centre for International Studies and Professor of International Relations at the University of Southampton. From 1982 to 1984 he served as the UK representative on the UN Secretary-General's Study Group on Conventional Disarmament. From 1993 to 1998 he was a member of the UN Secretary-General's Advisory Board for Disarmament Matters and since 1999 he has been an Advisor to the UK delegation to the NPT Review Conferences. Between 1987 and 2002 he served as Programme Director of the Programme for Promoting Nuclear Non-Proliferation (PPNN).

Professor Simpson gained his Ph.D for a thesis on UK weapons procurement and has been published widely in the areas of nuclear non-proliferation, the Non-Proliferation Treaty (NPT) and UK nuclear weapons policies.

GEORGE PERKOVICH is Vice President for Studies at the Carnegie Endowment for International Peace. He is an expert on US foreign policy, non-proliferation, security, global governance, non-governmental actors, India, Iran and Pakistan. He has also served as a speechwriter and foreign policy advisor to Senator Joe Biden (1989–1990).

Dr. Perkovich's personal research has focused on nuclear strategy and nonproliferation, with emphasis on South Asia. His book *India's Nuclear Bomb* received the Herbert Feis Award from the American Historical Association for outstanding work by an independent scholar, and the A.K. Coomaraswamy Prize from the Association for Asian Studies as an outstanding book on South Asia. Dr. Perkovich recently coauthored a major Carnegie report, *Universal Compliance: A Strategy for Nuclear Security*, a new blueprint for rethinking the international nuclear non-proliferation regime. His other publications include: *Iran Is Not an Island: A Strategy to Mobilize the Neighbors* (Policy Brief No. 34, February 2005); *Iran's Nuclear Challenge* (Carnegie Endowment Report, April 2003); and *WMD in Iraq: Evidence and Implications* (Carnegie Endowment Report, January 2004).

George Perkovich holds a BA degree from the University of California at Santa Cruz, an MA from Harvard University and a Ph.D from the University of Virginia.

JAMES H. NOYES is a Research Fellow at the Hoover Institution, Stanford University, specializing in Middle East Affairs. He is currently researching the influence of the Iraq War on evolving Middle East political conflicts. He was appointed Deputy Assistant Secretary of Defense for Near Eastern, African and South Asian Affairs in Washington, DC (1970–1976), and earlier served as Director for Syria with AMIDEAST in Damascus (1956–1958). He later became Representative of The Asia Foundation in Colombo (1965–1968), and was named Visiting Senior Fellow at the Institute of International Studies, University of California, Berkeley in 1997. After joining the Hoover Institution, he edited the Middle East section of the *Yearbook of Communist Affairs* (Hoover Institution Press) until the publication was halted with the collapse of the Soviet Union.

James Noyes' recent works include: *The US War on Terrorism: Impact on US-Arab Relations*, Emirates Lecture Series 49 (ECSSR, 2004); lectures at the Iraq Forum (2002 and 2003) sponsored by Oxford, Yale and Stanford in their AllLearn Internet course program for alumni; and an assessment of threats to US security interests in Southwest Asia over the next fifteen years prepared for the US Army's Strategic Studies Institute (June 2001). James Noyes' published work includes: "Does Washington Really Support Israel?" (*Foreign Policy*, Spring 1997) and "Assessing Prospects for Democracy in the Middle East," in *Change and Continuity in the*

Middle East: Conflict Resolution and Prospects for Peace (Macmillan Press, 1996). With M.E. Ahrari, he co-edited *The Persian Gulf after the Cold War* (Praeger Publishers, 1993) and contributed a chapter entitled: "American Perceptions of Iranian Threats to Gulf Security" in *The Gulf and International Security: The 1980s and Beyond* (Macmillan Press, 1989). His earlier volume, *The Clouded Lens: Persian Gulf Security and US Policy* (Hoover Institution Press, 1979 and revised edition, 1982), has been widely cited for its in-depth analysis of US strategic interests in Southwest Asia.

After receiving his BA from Yale University, James Noyes pursued his studies at Allahabad University, India, as a special student (1951–1952). He subsequently received his MA in Political Science from the University of California, Berkeley in 1954.

GEOFFREY ARONSON is Director of the Foundation for Middle East Peace in Washington, DC, where he writes the bi-monthly *Report on Israeli Settlements in the Occupied Territories* (www.fmep.org). The Report is distributed to the policymaking community in Washington and internationally to interested parties. He also serves on the board of Conflicts Forum, a London-based group which is interested in fostering discussions with Islamists in Palestine, Egypt and Lebanon.

Mr. Aronson writes widely on Middle East affairs for both journals and newspapers. Some of his recent articles have appeared in the *Journal of Palestine Studies*, *Financial Times* and *Daily Star*. He has just completed a paper funded by Canada's International Research and Development Center on the Palestinian Security Doctrine after Disengagement from Gaza. Earlier this year, he completed work on an extensive study of Israeli settlement assets and an assessment of how such assets might contribute towards meeting Palestinian housing and economic needs. During the Israeli disengagement from Gaza, he served as a consultant to the World Bank and the International Assessment Group.

SVERRE LODGAARD is Director of the Norwegian Institute of International Affairs. His areas of specialization include security policy; United Nations, NATO and European Union affairs; arms control and disarmament; security and development; peace operations; and Norwegian foreign affairs.

From 1992 to 1996, Mr. Lodgaard was the Director of the United Nations Institute for Disarmament Research (UNIDIR), at the UN Office in Geneva. Prior to this, Mr. Lodgaard was Director of the International Peace Research Institute in Oslo, and Director of European Security and Disarmament Studies at the Stockholm International Peace Research Institute (SIPRI). Mr. Lodgaard has held Research Fellowships at the International Peace Research Institute, the Department of Political Science at the University of Oslo and the Norwegian Research Council for Science and Humanities (NAVF).

In 1990, Mr. Lodgaard was awarded the Soka University Award of Highest Honor in Tokyo. He is a member of several academic institutions and councils such as the Bonn International Center for Conversion, the Pugwash Conferences on Science and World Affairs and the Editorial Committee of *Security Dialogue*. Sverre Lodgaard has a Magister degree in Political Science from the University of Oslo and supporting degrees in sociology and economics.

NOTES

Chapter 1

1. Report by the Director General, "Implementation of the NPT Safeguards Agreement in the Islamic Republic of Iran," International Atomic Energy Agency, 27 February 2006, GOV/2006/15, http://www.iaea.org/Publications/Documents/Board/2006/gov2006-15.pdf.

2. Seymour M. Hersh, "The Iran Plans," *The New Yorker*, April 17, 2006, http://www.newyorker.com/printables/fact/060417fa_fact.

3. *The National Security Strategy of the United States of America*, March 2006, The White House, 20, http://www.whitehouse.gov/nsc/nss/2006/nss2006.pdf.

4. This was the yield of the *Orange Herald Small* device exploded by the UK over Malden Island in the Pacific on May 31, 1957.

5. Report by the Director General, "Implementation of the NPT Safeguards Agreement in the Islamic Republic of Iran," International Atomic Energy Agency, 27 February 2006, GOV/2006/15, http://www.iaea.org/Publications/Documents/Board/2006/gov2006-15.pdf.

6. Ibid.

7. Unclassified Summary of a National Intelligence Estimate, *Foreign Missile Developments and the Ballistic Missile Threat Through 2015* (December 2001), http://www.fas.org/irp/nic/bmthreat-2015.htm.

8. *Current and Projected National Security Threats to the United States*, Vice Admiral Lowell E. Jacoby, US Navy, Director, DIA, Statement for the Record, Senate Select Committee on Intelligence, February 11, 2003, http://www.fas.org/irp/congress/2003_hr/021103jacoby.html.

9. Everts, S., "The EU and Iran: how to make conditional engagement work," Centre for European Reform, Policy Brief, 2003, 2, http://www.cer.org.uk/pdf/policybrief_eu_iran.pdf.

10. David Albright and Corey Hinderstein, "Iran, player or rogue?" *Bulletin of Atomic Scientists*, vol. 59, no. 5, 57–8, http://www.thebulletin.org/print.php?art_ofn=so03albright.

11. David Albright and Corey Hinderstein, *Iran's Next Steps: Final Tests and the Construction of a Uranium Enrichment Plant*, Institute for Science and International Security (ISIS), January 12, 2006, http://www.isis-online.org/publications/iran/irancascade.pdf.

12. *Current and Projected National Security Threats to the United States*, Vice Admiral Lowell E. Jacoby, US Navy, Director, DIA, Statement for the Record, Senate Armed Services Committee, March 17, 2005, 10, http://www.dia.mil/publicaffairs/Testimonies/20050317_DR_Jacoby_WWT_SASC_SFR-U-Final.pdf.

13. Sokolski, H., "Getting Ready for a Nuclear-ready Iran: Report of the NPEC Working Group" in Henry Sokolski and Patrick Clawson (eds), *Getting Ready for a Nuclear-ready Iran*, Strategic Studies Institute Publication (October 2005), http://www.strategicstudiesinstitute.army.mil/pdffiles/PUB629.pdf.

14. Joseph Cirincione, Carnegie Issue Brief, *No Military Options*, January 19, 2006, http://www.carnegieendowment.org/publications/index.cfm?fa=print&id=17922.

15. David Albright and Corey Hinderstein, *The Clock is Ticking, But How Fast?* Institute for Science and International Security (ISIS), March 27, 2006, 4–7, http://www.isis-online.org/publications/iran/clock ticking.pdf.

16. Ibid., indeed one sub-option puts the conversion time at "about one to two months," 7.

17. Report by the Director General, "Implementation of the NPT Safeguards Agreement in the Islamic Republic of Iran," International Atomic Energy Agency, September 2, 2005, GOV/2005/67, http://www.iaea.org/Publications/Documents/Board/2005/gov2005-67.pdf.

18. David Albright and Corey Hinderstein, "Countdown to Showdown," *Bulletin of the Atomic Scientists*, November/December 2004, 70.

19. Mark Fitzpatrick, IISS Non-proliferation analyst, former US State Department Undersecretary for Nonproliferation, *ABC Online*, "Britain claims Iran in pursuit of nuclear technologies," January 5, 2006, http://www.abc.net.au/cgi-bin/common/printfriendly.pl?http://www.abc.net.au/am/content/2006/s1542219.htm.

Chapter 2

1. *Kayhan*, July 23, 2005, 12.

2. Ibid., 7.

3. Ibid.

4. My colleagues and I at the Carnegie Endowment for International Peace have sketched steps that would be more helpful in: *Universal Compliance: A Strategy for Nuclear Security*.

5. My colleague Pierre Goldschmidt has devised such a fuel-service mechanism.

6. *Kayhan*, op. cit.

Chapter 3

1. The Gulf Cooperation Council (GCC), founded in 1981, is the consultative body of the Arab Gulf states of Bahrain, Kuwait, Oman, Qatar, Saudi Arabia and the United Arab Emirates.

2. "Security Turnaround: an Interview with Nawaf Obaid," Bitterlemons-international.org, edition 2, vol. 4, January 19, 2006.

3. Ibid.

4. Anthony H. Cordesman and Nawaf Obaid, *National Security in Saudi Arabia: Threats, Responses, and Challenges* (Westport, CT: Praeger Security International/CSIS, 2005), 57.

5. *Frankfurter Allgemeine Zeitung*, September 13, 2004.

6. Ibid., 51–2.

7. Middle East Media Research Institute (MEMRI), clip no. 1109, March 21, 2006, Channel 2, Iranian TV, http://www.memritv.org/Transcript.asp?P1=1109.

8. Reuters, "Iran Fires Missile That Can Evade Radar: TV," *The New York Times,* March 31, 2006.

9. Middle East Media Research Institute (MEMRI), clip no. 1106, April 4, 2006, Jaam-e Jam 3 TV, http://www.memritv.org/Transcript.asp?P1=1106.

10. Middle East Media Research Institute (MEMRI), clip no. 1102, April 4, 2006, Iranian TV, Channel 1, http://www.memritv.org/Transcript.asp?P1=1102.

11. Associated Press, "Iran: High-speed underwater missile test-fired," CNN.com, April 2, 2006.

12. William J. Broad, and David E. Sanger, "Iran Joins the Space Club, but Why?" *The New York Times,* April 4, 2006.

13. Ibid.

14. Diana Elias, "Iran's Nukes Concern Some Arab Countries," http://news.yahoo.com/s/ap/20060322/ap.

15. Middle East Media Research Institute (MEMRI), clip no. 1101, March 31, 2006, Al-Arabiya TV, http://www.memritv.org/Transcript.asp?P1=1101.

16. Ibid.

17. Jonathan Wright, "Many Arabs favor nuclear Iran," http://www.news.yahoo.com/s/nm/20060418/wl_nm/nuclear_iran_arabs_dc_1.

18. Elias, op. cit.

19. The Emirates Center for Strategic Studies and Research (ECSSR), "Founded Apprehension Over Iran's Nuclear Program," *Akhbar Al-Sa'ah*, vol.12, no. 3219, December 19, 2005.

20. Cordesman and Obaid, op. cit., 249–50.

21. Middle East Media Research Institute (MEMRI), clip no. 1069, March 6, 2006, Al-Rai TV, http://www.memritv.org/Transcript.asp?P1=1069.

22. Details of the transaction and its ultimate impact on US–Saudi relations are provided in: Rachel Bronson, *Thicker Than Oil: America's Uneasy Partnership with Saudi Arabia* (New York, Oxford University Press, 2006), 188–90.

23. John M. Goshko and Don Oberdorfer, "Chinese Sell Saudis Missiles Capable of Covering the Middle East," *Washington Post*, March 18, 1988.

24. See Anthony H. Cordesman and Nawaf Obaid, op. cit., 137–139.

25. Ibid., as cited from *Jane's Sentinel Security Assessment,* August 27, 2004.

26. Middle East Media Research Institute (MEMRI), Special Dispatch Series, no. 1130, March 30, 2006, ASWAT (the Independent Iraqi News Agency).

27. Michael Theodoulou, "Saudis plan to fence off boarder with chaos," *The Times*, April 10, 2006, http://www.timesonline.co.uk/printFriendly/0,,1-23-2126835-23,00.html.

28. David E. Sanger, "Why Not a Strike on Iran?" *The New York Times*, January 22, 2006.

29. Paul Rogers, *IRAN: Consequences of a War*, www.oxfordresearchgroup.org.uk, February 2006.

30. US Energy Information Administration (EIA), *Iran*, Country Analysis Briefs, January 2006, www.eia.doe.gov.

31. Editorial, *Wall Street Journal*, February 3, 2006.

32. Dafna Linzer, "Pro-Israel Group Criticizes White House Policy on Iran," *Washington Post*, December 25, 2005.

33. US Energy Information Administration (EIA), op. cit.

34. Ibid.

35. Middle East Media Research Institute (MEMRI), clip no. 1126, 3/28/06, Internet Daily, www.roozonline.com, March 19, 2006, http://www.memritv.org/Transcript.asp?P1=1126.

36. Ray Takeyh, "A Profile in Defiance: Being Mahmoud Ahmadinejad," *The National Interest*, Spring 2006, no. 83, 16–21.

37. *The National Strategy for Victory in Iraq*, http://www.whitehouse.gov/infocus/iraq/iraq_national_strategy_20051130.pdf.

Chapter 4

1. Foundation for Middle East Peace, *Special Report: The Uncertainties of Peace: Regional Implications of Israeli–Arab Rapprochement*, vol. SR no. 9, Spring 2000.

2. Johnathan Marshall, Peter Dale Scott and Jane Hunter, *The Iran Contra Connection, Secret Teams and Covert Operations in the Reagan Era*, (Boston, MA: South End Press, 1987).

3. *Al Wasat*, December 2, 1993.

4. Ibid.

5. Ibid.

6. *Al Wasat*, January 3, 1996.

7. "Conceptualizing U.S. Strategy in the Middle East," Anthony Lake, Assistant to the President for National Security Affairs, Washington Institute for Near East Policy, Soref Symposium, 1994.

8. "Beyond the Pale," Mike Moore, http://www.thebulletin.org/article.php?art_ofn=jf99moore_045; *Al Wasat*, May 1, 1999.

9. *Davar*, April 17, 1992.

10. *Ha'aretz*, April 27, 1992.

11. *Al Wasat*, May 1, 1999.

12. *Al Wasat*, July 10, 1993.

13. *Al Wasat*, April 28, 1997.

14. *Yedioth Ahronoth*, September 29, 1991; *Ma'ariv*, April 17, 1992.

15. *Ha'aretz*, May 20, 1998.

16. At 2 pm (local time) on October 6, 1973, Egypt and Syria attacked Israel in a coordinated surprise attack, starting the Yom Kippur War. Caught with only their standing forces on duty – and these at a low level of readiness – the Israeli front lines were overrun. By early afternoon on October 7, no defensive forces were left in the southern Golan Heights

and Syrian forces had reached the edge of the plateau, within sight of the Jordan River. It has been widely reported that this crisis brought Israel to its first nuclear alert. This resulted in the Jericho missiles at Hirbat Zachariah and the nuclear strike F-4 Phantoms at Tel Nof being armed and prepared for action against Syrian and Egyptian targets. US Secretary of State Henry Kissinger was apparently notified of this alert several hours later on the morning of October 9, which helped motivate a US decision to promptly open a resupply pipeline to Israel (Israeli aircraft began picking up supplies that day, the first US flights arrived on October 14), "Israel's Nuclear Weapons Program," December 10, 1997, http://nuclearweaponarchive.org/Israel/Isrhist.html.

17. "The Link between Peace and the Atom," Reuven Pedatzur, *Ha'aretz*, January 4, 1996.

18. "This is the test of the United States in the region," noted Prime Minister Yitzhak Rabin shortly before his election in July 1992, "Whether it will manage to curb and prevent the proliferation of nuclear weapons … Only the US can do this and I hope that it will."

19. Israeli Chief of Staff, Gen. Dan Halutz, *Arutz 7*, January 17, 2006.

20. "After the Hamas Victory: The Increasing Importance of Israel's Strategic Barrier in the Jordan Valley," Dore Gold, Jerusalem Center for Public Affairs, February 7, 2006.

21. *Ha'aretz*, January 23, 2006.

22. See, for example, annual assessment by Mossad chief Meir Dagan before the Knesset, in *Ha'aretz*, December 30, 2005.

23. US president George W. Bush at the City Club of Cleveland, March 20, 2006.

24. *Al Wasat*, April 2000.

25. The statement notes: "Recognizing that a solution to the Iranian issue would contribute to global nonproliferation efforts and to realizing the objective of a Middle East free of weapons of mass destruction, including their means of delivery," http://www.globalsecurity.org/wmd/library/report/2006/iran_iaea_gov2006–14_4feb06.htm.

26. "Hidden Agenda: US–Israeli Relations and the Nuclear Question," Geoffrey Aronson, *Middle East Journal*, Autumn 1992.

27. "U.S. Spent $1.9 Million to Aid Fateh in the Elections," Steven Erlanger, *The New York Times*, January 23, 2006.

Chapter 5

1. *The National Security Strategy of the United States of America*, September 2002, http://www.whitehouse.gov/nsc/nss.pdf.

2. *The National Security Strategy of the United States of America*, March 2006, http://www.whitehouse.gov/nsc/nss/2006/nss.pdf.

3. As quoted in Ted Koppel, "Will Fight for Oil," *The New York Times*, February 24, 2006.

4. Ted Koppel, op. cit.

5. Seymour M. Hersh, "The Iran Plans," *The New Yorker*, April 17, 2006, http://www.newyorker.com/printables/fact/060417fa_fact.

6. For declassified excerpts of the "Nuclear Posture Review" submitted to Congress on December 31, 2001, see http://www.globalsecurity.org.wmd/library/policy/dod/npr.htm.

7. Seymour Hersh says that the political leadership wants to keep nuclear options in the war plans while military leaders warn against it. See Seymour M. Hersh, op. cit.

8. Seymour M. Hersh, op. cit. Hersh makes a comprehensive account of US planning in this respect. The accuracy of it is open to doubt, however.

9. "Response of the Islamic Republic of Iran to the Framework Agreement proposed by the EU-3/EU," printed in INFCIRC/651, IAEA, August 8, 2005.

10. "Framework for a Long-Term Agreement between the Islamic Republic of Iran and France, Germany and the United Kingdom, with the Support of the High Representative of the European Union," paragraph 24, printed in INFCIRC/651, IAEA, August 8, 2005.

11. Ibid., paragraph 19b.

12. "Response of the Islamic Republic of Iran," op. cit.

13. "Bush Hints at Military Option for Iran," *Herald Sun*, August 13, 2005.

14. "Germany Attacks US on Iran Threat," BBC News, August 13, 2005.

15. Bruno Pellaud said: "A joint nuclear facility with multinational staff puts all participants under a greater scrutiny from peers and partners, a fact that strengthens non-proliferation and security ... It's difficult to play games if you have multinationals at a site."

16. "Multinational Facilities may Solve Iranian Nuclear Stalemate," *Jane's Intelligence Review*, no. 4, April 2006.

17. President Mahmoud Ahmadinejad in a speech before the United Nations, New York, September 17, 2005.

18. South Africa, when relinquishing its nuclear weapons, made virtually everything accessible to the Agency at once.

19. In view of Iran's past pattern of concealment and disinformation; in the face of major power claims based on national intelligence and

information protected by national red tape and therefore not available to it, the Agency has obviously taken care not to conclude prematurely. A positive conclusion – however guarded – proven erroneous by later findings, could jeopardize its reputation.

20. The "inalienable right" to peaceful utilization of nuclear energy inscribed in Article IV of the NPT does not apply irrespective of compliance or non-compliance with safeguards obligations.

21. The P-2 rotors are based on maraging steel and can work at twice the speed of P-1 rotors, which are made of aluminum. The Iranian government says that it did no work on the P-2 until 2002, when the design information was given to a small firm in Tehran. In a short period of time this firm developed a modified version so expeditiously that the IAEA finds the story unrealistic unless the process was assisted by someone else. The Agency therefore concluded that: "The reasons given by Iran for the apparent gap between 1995 and 2002, however, do not provide sufficient assurance that there were no related activities carried out in that period," IAEA, GOV/2004/83, November 15, 2004, 11.

22. This expression was used in conjunction with the summit meeting in October 2004, when long-standing border issues were settled and Moscow and Beijing agreed to hold joint military exercises in 2005, for the first time since 1958.

23. Some observers believe that the US has deliberately escalated the nuclear conflict with Iran in order to give the Iranians incentives to cooperate in settling problems in Iraq—problems which may be intractable without Iranian assistance.

24. Briefing on the Iranian nuclear issue, R. Nicholas Burns, Robert Joseph, April 21, 2006. Others would say that the point of no return is when Iran has produced enough fissile material for a nuclear weapon, or the point at which it has been successfully weaponized.

BIBLIOGRAPHY

Ahmadinejad, President Mahmoud. Speech before the United Nations, New York, September 17, 2005.

Albright, David and Corey Hinderstein. "Iran, player or rogue?" *Bulletin of the Atomic Scientists*, vol. 59, no. 5, 2003 (http://www.thebulletin.org/print.php?art_ofn=so03albright).

Albright, David and Corey Hinderstein. "Countdown to Showdown." *Bulletin of the Atomic Scientists* (November/December, 2004).

Albright, David and Corey Hinderstein. *Iran's Next Steps: Final Tests and the Construction of a Uranium Enrichment Plant.* Institute for Science and International Security (ISIS), January 12, 2006 (http://www.isis-online.org/publications/iran/irancascade.pdf).

Albright, David and Corey Hinderstein. *The Clock is Ticking, But How Fast?* Institute for Science and International Security (ISIS), March 27, 2006 (http://www.isis-online.org/publications/iran/clockticking.pdf).

Al Wasat, April 2000.

—May 1, 1999.

—April 28, 1997.

—January 3, 1996.

—December 2, 1993.

—July 10, 1993.

Aronson, Geoffrey. "Hidden Agenda: US–Israeli Relations and the Nuclear Question." *Middle East Journal*, Autumn 1992.

Associated Press (AP). "Iran: High-speed underwater missile test-fired." CNN.com, April 2, 2006.

Broad, William J. and David E. Sanger. "Iran Joins the Space Club, but Why?" *The New York Times*, April 4, 2006.

Bronson, Rachel. *Thicker Than Oil: America's Uneasy Partnership with Saudi Arabia* (New York, Oxford University Press, 2006).

Burns, R. Nicholas and Robert Joseph. Briefing on the Iranian nuclear issue, April 21, 2006.

"Bush Hints at Military Option for Iran." *Herald Sun*, August 13, 2005.

Bush, President George W. Speech at the City Club of Cleveland, March 20, 2006.

Cirincione, Joseph. *No Military Options.* Carnegie Issue Brief, January 19, 2006 (http://www.carnegieendowment.org/publications/index.cfm?fa=print&id=17922).

Cordesman, Anthony H. and Nawaf Obaid. *National Security in Saudi Arabia: Threats, Responses and Challenges* (Westport, CT: Praeger Security International/CSIS, 2005).

Davar, April 17, 1992.

Elias, Diana. "Iran's Nukes Concern Some Arab Countries" (http://www.yahoonews.com/s/ap/20060322/ap).

Erlanger, Steven. "US Spent $1.9 Million to Aid Fateh in the Elections." *The New York Times*, January 23, 2006.

Everts, S. "The EU and Iran: How to make Conditional Engagement work." Centre for European Reform, Policy Brief, 2003 (http://www.cer.org.uk/pdf/policybrief_eu_iran.pdf).

Fitzpatrick, Mark. "Britain claims Iran in pursuit of nuclear technologies." *ABC Online*, January 5, 2006 (http://www.abc.net.au/cgi-bin/common/printfriendly.pl?http://www.abc.net.au/am/content/2006/s1542219.htm).

Foreign Missile Developments and the Ballistic Missile Threat through 2015. Unclassified Summary of a National Intelligence Estimate, December 2001 (http://www.fas.org/irp/nic/bmthreat-2015.htm).

Foundation for Middle East Peace. *Special Report: The Uncertainties of Peace: Regional Implications of Israeli–Arab Rapprochement.* Vol. SR no. 9, Spring 2000.

Frankfurter Allgemeine Zeitung, September 13, 2004.

"Germany Attacks US on Iran Threat." BBC News, August 13, 2005.

Gold, Dore. *After the Hamas Victory: The Increasing Importance of Israel's Strategic Barrier in the Jordan Valley.* Jerusalem Center for Public Affairs, February 7, 2006.

Goshko, John M. and Don Oberdorfer. "Chinese Sell Saudis Missiles Capable of Covering the Middle East." *Washington Post*, March 18, 1988.

Ha'aretz, January 23, 2006.

— May 20, 1998.

— April 27, 1992.

Halutz, Gen. Dan. *Arutz 7*, January 17, 2006.

Hersh, Seymour M. "The Iran Plans." *The New Yorker*, April 17, 2006 (http://www.newyorker.com/printables/fact/060417fa_fact).

International Atomic Energy Agency (IAEA). *Framework for a Long-Term Agreement between the Islamic Republic of Iran and France, Germany and the United Kingdom, with the Support of the High Representative of the European Union.* INFCIRC/651, August 8, 2005.

International Atomic Energy Agency (IAEA). *Implementation of the NPT Safeguards Agreement in the Islamic Republic of Iran.* Report by the Director General. GOV/2006/15, February 27, 2006 (http://www.iaea.org/Publications/Documents/Board/2006/gov2006-15.pdf).

International Atomic Energy Agency (IAEA). *Implementation of the NPT Safeguards Agreement in the Islamic Republic of Iran*. Report by the Director General. GOV/2005/67, September 2, 2005 (http://www.iaea.org/Publications/Documents/Board/2005/gov2005-67.pdf).

International Atomic Energy Agency (IAEA). *Implementation of the NPT Safeguards Agreement in the Islamic Republic of Iran*. Report by the Director General. GOV/2004/83, November 15, 2004.

International Atomic Energy Agency (IAEA). *Response of the Islamic Republic of Iran to the Framework Agreement proposed by the EU-3/EU*. INFCIRC/651, August 8, 2005.

"Israel's Nuclear Weapons Program." December 10, 1997, (http://nuclearweaponarchive.org/Israel/Isrhist.html).

Jacoby, Vice Admiral Lowell E. *Current and Projected National Security Threats to the United States*. Statement for the Record, Senate Select Committee on Intelligence, February 11, 2003 (http://www.fas.org/irp/congress/2003_hr/021103jacoby.html).

Jacoby, Vice Admiral Lowell E. *Current and Projected National Security Threats to the United States*. Statement for the Record, Senate Armed Services Committee, March 17, 2005 (http://www.dia.mil/publicaffairs/Testimonies/20050317_DR_Jacoby_WWT_SASC_SFR-U-Final.pdf).

Jane's Sentinel Security Assessment, August 27, 2004.

Kayhan, July 23, 2005.

Koppel, Ted. "Will Fight for Oil." *The New York Times*, February 24, 2006.

Lake, Anthony. "Conceptualizing U.S. Strategy in the Middle East." Washington Institute for Near East Policy, Soref Symposium, 1994.

Linzer, Dafna. "Pro-Israel Group Criticizes White House Policy on Iran." *Washington Post*, December 25, 2005.

Ma'ariv, April 17, 1992.

Marshall, Johnathan, Peter Dale Scott and Jane Hunter. *The Iran Contra Connection, Secret Teams and Covert Operations in the Reagan Era* (Boston, MA: South End Press, 1987).

Middle East Media Research Institute (MEMRI). Clip no. 1126, 3/28/06, Internet Daily, www.roozonline.com, March 19, 2006 (http://www.memritv.org/Transcript.asp?P1=1126).

Middle East Media Research Institute (MEMRI). Clip no. 1109, March 21, 2006, Channel 2, Iranian TV (http://www.memritv.org/Transcript.asp?P1=1109).

Middle East Media Research Institute (MEMRI). Clip no. 1106, April 4, 2006, Jaam-e Jam 3 TV (http://www.memritv.org/Transcript.asp?P1=1106).

Middle East Media Research Institute (MEMRI). Clip no. 1102, April 4, 2006, Iranian TV, Channel 1 (http://www.memritv.org/Transcript.asp?P1=1102).

Middle East Media Research Institute (MEMRI). Clip no. 1101, March 31, 2006, Al-Arabiya TV (http://www.memritv.org/Transcript.asp?P1=1101).

Middle East Media Research Institute (MEMRI). Clip no. 1069, March 6, 2006, Al-Rai TV (http://www.memritv.org/Transcript.asp?P1=1069).

Middle East Media Research Institute (MEMRI). Special Dispatch Series, no. 1130, March 30, 2006, ASWAT (the Independent Iraqi News Agency).

Moore, Mike. "Beyond the Pale" (http://www.thebulletin.org/article.php?art_ofn=jf99moore_045).

"Multinational Facilities may Solve Iranian Nuclear Stalemate." *Jane's Intelligence Review*, no. 4, April 2006.

"Nuclear Posture Review." Submitted to Congress on December 31, 2001 (http://www.globalsecurity.org/wmd/library/policy/dod/npr.htm).

Pedatzur, Reuven. "The Link between Peace and the Atom." *Ha'aretz*, January 4, 1996.

Reuters. "Iran Fires Missile That Can Evade Radar: TV." *The New York Times,* March 31, 2006.

Rogers, Paul. *IRAN: Consequences of a War*. Oxford Research Group, February 2006 (http://www.oxfordresearch group.org.uk/publications/briefings/IranConsequences.htm).

Sanger, David E. "Why Not a Strike on Iran?" *The New York Times,* January 22, 2006.

"Security Turnaround: an Interview with Nawaf Obaid." Bitterlemons-international.org, edition 2, vol. 4, January 19, 2006.

Sokolski, Henry. "Getting Ready for a Nuclear-ready Iran: Report of the NPEC Working Group," in Henry Sokolski and Patrick Clawson (eds), *Getting Ready for a Nuclear-ready Iran.* Strategic Studies Institute Publication, October 2005 (http://www.strategicstudiesinstitute.army.mil/pdffiles/PUB629.pdf).

Takeyh, Ray. "A Profile in Defiance: Being Mahmoud Ahmadinejad." *The National Interest*, no. 83, Spring 2006.

The Emirates Center for Strategic Studies and Research (ECSSR). "Founded Apprehension Over Iran's Nuclear Program." *Akhbar Al-Sa'ah*, vol. 12, no. 3219, December 19, 2005.

The National Security Strategy of the United States of America. The White House, Washington, DC, September 2002 (http://www.whitehouse.gov/nsc/nss.pdf).

The National Security Strategy of the United States of America. The White House, Washington, DC, March 2006 (http://www.whitehouse.gov/nsc/nss/2006/nss2006.pdf).

The National Strategy for Victory in Iraq. The White House, Washington, DC, 2005 (http://www.whitehouse.gov/infocus/iraq/iraq_national_strategy_20051130.pdf).

Theodoulou, Michael. "Saudis plan to fence off boarder with chaos." *The Times*, April 10, 2006 (http://www.timesonline.co.uk/printFriendly/0,,1-23-2126835-23,00.html).

US Energy Information Administration (EIA). *Iran*, Country Analysis Briefs, January 2006 (http://www.eia.doe.gov).

Wall Street Journal, Editorial, February 3, 2006.

Wright, Jonathan. "Many Arabs favor nuclear Iran" (http://www.news.yahoo.com/s/nm/20060418/wl_nm/nuclear_iran_arabs_dc_1).

Yedioth Ahronoth, September 29, 1991.

INDEX

A
Abassi, Hassan 86
Abdullah, King of Saudi Arabia 65
Abqaiq 78
Afghanistan
    Iranian links 122
    lack of US political success 69
    Taliban removal 84
    US forces in 84, 116
Ahmadinejad, Mahmoud
    2005 World Summit speech 46
    aggressive stance 37, 46–7, 53–4, 61, 66, 69–70
    centrifuge research announcement 133
    and "Council of Heads" 44
    distrust of US/West 87
    electoral victory 37, 43, 86
    hopes to deal with East 51
    on Iran as regional leader 65
    is international dispute deliberate? 66
    nuclear limitations flouted 37, 46–8
    "opportunity" provided by 50, 52
    shared ownership idea 128
    threats against Israel 81, 83
AIPAC *see* American Israel Public Affairs Committee
Al-Assad, Bashar 114
Al Bu Ainnain, Brig. General Khalid 77
Al-Din, Sayyed Ayad Jamal 79
Al-Faraj, Sami 73
Al-Hasa region 73
Al Qaeda 88
    Iraq stronghold fears 89
    as Saudi threat 65, 78, 80
Al-Rubei, Ahmad 75
Al-Sha'er, Sawsan 73
Al-Shirian, Dawood 75
Al-Zarkawi, Abu Musab 78, 88

Albright, David 30–1, 32
America *see* United States
American Israel Public Affairs Committee (AIPAC) 83
Annan, Kofi 46
Arab–Israeli conflict
    Bush administration view of 103
    intifada ends rapprochement (2000) 103
    Israel's stabilization attempts 95–6
    Rabin's peace hopes 97
Arafat, Yasser 96
Arak facility 44
Armenia 86
Azerbaijan 86

B
Baathist party
    as check on Iran 84
    as jihadist barrier 72–3
    secularism of 86
Bahrain
    as CTBT signatory 22
    Shiite majority 73, 88
    as US base 82, 89
Barak, Ehud 99, 100, 102
Beilin, Yossi 109–10
Beirut Marine barracks bombing 64
Belarus 114
Ben-Gurion, David 93
Blue Danube 26
Britain
    first bomb design 26
    see also United Kingdom
Burma 114
Bush administration
    and Arab–Israeli peace 103
    condemnation of Iran 100
    criticized over Iran 83

democracy agenda 53
nuclear aid for Israel 108
possible use of nuclear arms 120
unilateral foreign policy 135
Bush, George 110
Bush, George W.
"Bush doctrine" 113
Middle East
arms control proposal (1991) 138
focus on 113
threat of force against Iran (2005) 126, 127
hopes for victory in Iraq 88
Bushehr reactor
accident fears 73, 75
Russian building of 30
Russian fuel for 22

C
Canada 130
Carter, Jimmy 116
centrifuge technology *see under* uranium enrichment
Chechnya
Chinese/Russian/Iranian position 135
rebellion 86
Cheney, Dick 116
China 50, 51, 55, 57
Iranian investments/agreements 134
Israeli pressure on 97
limited support for Iran 68
nuclear weapons 33, 60
pressure on Iran 64, 66
Sino-Russian relations 134–5
strong Iranian ties 134
US unilateralism rejected 135
Christian Armenia *see* Armenia
Cirincione, Joseph 32
Clinton, William ("Bill") 100
Comprehensive Test Ban Treaty (CTBT) 21
signatories of 22
status of 22–3
Cuba 114

D
Dehghani, Mohammad Rahim 71
delivery systems
induction of 25
missiles as 23, 24–5
possible Iranian choice of 25–6
restrictions on use 21
deuterium 17
Dimona reactor 27, 138
dual-use options 12, 13, 15, 29

E
ECSSR *see* Emirates Center for Strategic Studies and Research
Egypt
as CTBT signatory 22
as possible nuclear developer 77, 106
Israel's nuclear opposition to 106, 111
mixed reactions to Iran 74–5
peace with Israel 104
WMD-free Middle East proposition 137
el-Erian, Essam 74
ElBaradei, Mohamed 47
Emirates Center for Strategic Studies and Research (ECSSR) 75
Esfahan facility
operations begin (2005) 45, 61
work continues during suspension 44–5
Ethiopia 93
EU-3
Iran
enrichment ban (2006) 131
fissile fuel fears 40
lack of inducements 45
nuclear fuel imports plan 40
unwilling to penalize 45
voluntary nuclear suspension (2003) 39–41, 40–1, 42

voluntary suspension resumed
(2004) 43
suspension terms breached
(2004) 42–3
EU-3/EU Framework for a Long-Term
Agreement
fuel supply offer 124–5, 131
Iran rejects offer 127
Iran's right to nuclear power
program 125
Iran's security not addressed 125–6
United States attitude/concessions
126–7
Eurodif plant 128
Europe 50, 57
Israeli pressure on 97
nuclear enrichment plants 128
pressure on Iran 66
export controls
global guidelines 23
listed nuclear items 23, 24–6
missile systems 23

F
Fadavi, General Ali 71–2
Faw Peninsula 63
fissile materials
amount required 19–20
description of 17
fission (atomic) devices
definition of 16
initiation methods
gun 19, 20
implosion 19–20
materials 17, 19–20
France
dual-use options 29
as EU-3 member 39
Iran enrichment opposition 131
fusion devices *see* thermonuclear
(hydrogen) devices

G
gas: Iran's resources 67
GCC *see* Gulf Cooperation Council
(GCC)

Geneva (2000) 103
Germany
against attack on Iran 127–8
as EU-3 member 39
and Iranian enrichment 131
Israeli pressure on 97
nuclear capability 15
global community *see* international
community
Golan Heights 105
Gulf Cooperation Council (GCC) 51
defense capabilities 77
Iran's hegemony claims 65, 67, 71
Israeli "double standard" problem 74
nuclear-free region hopes 91
reactions to nuclear Iran 75
reactor accident fears 73
security
Al Qaeda threat 78, 80, 89
continuing US presence 89
effect of US strike on Iran 81–2
Iran's influence on Shiite
minority 73, 78
Iraq's disintegration scenario
87–9
Iraqi events threat 64, 78–80, 87
Islamic "modernists" vs.
traditionalists 64
spectrum of issues 64
unreliable intelligence 80
terrorism 64, 65, 78
threat of unstable Iran 64, 65–6,
70, 80
US/Israeli Iran strike
repercussions 80
Gulf region
military miscalculations 63
US military dominance 116
Gulf War 102, 105, 108, 110

H
Ha'aretz 98
Habib, Mohammed 74–5

Hague Code of Conduct against
  Ballistic Missile Proliferation
  (HCoC) 23
Hamas 88
  Iranian support 66
  Palestinian victory 122
HCoC *see* Hague Code of Conduct
  against Ballistic Missile Proliferation
heavy water 18
HEU *see* high-enriched uranium
  (HEU)
Hezbollah 85, 88, 104
  attacks on US bases 78
  divided loyalties 91
  Iranian support for 66, 108–9
  threat to Israel 104, 122
high-enriched uranium (HEU) 31, 32,
  33, 34, 35
Hinderstein, Corey 30–1, 32
Hiroshima 19, 20
Hussein, Saddam
  attack on Iran (1980) 63
  collapse of forces 65
  jihadist resistance 72–3
  negative result of removal 104
  nuclear development 94
  West's WMD claims 68
  see also Baathist party
hybrid devices 17, 28

I
IAEA *see* International Atomic Energy
  Agency (IAEA)
India 50, 51, 57, 64
  covert nuclear development 80
  early nuclear activity 26
  nuclear deterrent status 27, 66
  Pakistan's nuclear threat 59
  pressure on Iran 66
Indo-Pak weapons tests (1998) 59
International Atomic Energy Agency
  (IAEA)
  clandestine activities addressed 13

export control list 23
Iran
  fissile material manufacture
    (2003–) 29–30
  inspections access denied 47
  inspections allowed 129
  limited enrichment suggestion
    131
  non-compliance declaration
    (2005) 39
  nuclear activities revealed
    (2002) 12, 37, 38
  nuclear fuel supply scheme 128
  reaction to Natanz resumption
    (2006) 47
  report to UN Security Council
    (2006) 133
  unwilling to penalize 45
multinational fuel supply scheme 128
non-proliferation safeguards 13
Nuclear-Free Zone (NFZ) call 109
safeguards inspections 22
WMD-free Middle East proposed
  (2006) 137
international community
  enrichment moratorium plans 130
  Iran
    disagreements among major
      powers 123–4
    call to control all Iranian
      weapons 109
    enrichment ban (2006) 131
    general diplomatic initiatives
      127–32
    lack of confidence in 12
    need for united front 50–1, 52,
      54, 57–8, 60–1
    nuclear-armed Iran fears 12,
      18, 25, 35–6, 38, 48
    possible anti-nuclear strategies
      48–60, 106–8
    violations discovered (2002) 37
  Iran: possible strategies

build on Ahmadinejad's
    aggression 50, 52
change domestic debate
    dynamic 48–9, 50
create Middle East WMD-free
    zone 51–2
create compelling global
    coalition 50–1, 52, 54, 57–8
multilateral fuel cycle option
    128–9, 131
offer cost-beneficial nuclear
    fuel 52–3
offer regime security 53–4
offer rewards vs. penalties 50,
    52–3
the way ahead 132
see also sanctions
nuclear proliferation
    barriers to 21–2, 23
    non-proliferation doubts 12–13
nuclear testing moratorium 23
see also EU-3; EU-3/EU
intifada (2000) 103
Iran
    as CTBT signatory 22
    economy 85
    ethnic minority split 85–6
    fears of Iraqi WMDs 68
    foreign policy
        "brinksmanship" policies 60
        confrontation with West 66, 70
        as threat to GCC 64, 65–6, 70
        Gulf leadership claim 65, 67, 71
        Israel's destruction urged 46,
            68, 69, 90, 116–17
        nuclear weapons as a threat
            59–60
        pragmatism vs. religion 86
        regional domination hopes 90–1
        support for terrorism 66, 108–9
    gas resources 67
    global lack of confidence in 12
    government unpopularity 91

historic Persian roots 11
India–Pakistan nuclear fears 66
industrial state hopes 11
Islamic Revolution 94
Israel
    destruction urged 46, 68, 69,
        90, 116–17
    pre-revolution cooperation 93, 94
    rapprochement potential 109–10
    reasons for Israel's hostility 95
military
    belligerent equipment claims
        70–2
    outmoded equipment 68–9
    possible GCC reprisals 81–2
    regional superiority 65
    vulnerability against US 84–5
military action against: scenarios
    consequences of 121–3
    "contain-and-deter" 58, 60
    concurrent covert operations 121
    intelligence problem 57
    likely targets of 119–21
    possibility of 115, 117–18
    what happens next? 57
    will Israel or US attack? 119–21
missiles
    case for deterrent use 69
    "completely new" missiles 71
    nuclear-compatible plans 25
    planned system 48–9
    Shahab improvements 70
    support for Hezbollah 108–9
    underwater missiles 71–2
nuclear activities
    1980s/1990s global suspicions 38
    back-down unlikely 135
    "entirely peaceful" claims 12,
        41, 67
    IAEA non-compliance
        declaration (2005) 39
    IAEA photographic evidence
        (2002) 38

indigenous fuel aim 40–1
"red line" fixed by Council of Heads (2005) 44
regional reasons for 68
Russian offer rejected (2005) 46
nuclear inspections
  activities revealed by IAEA (2002) 12, 37, 38
  Additional Protocol rejected 129
  IAEA inspections allowed 129
nuclear scenarios
  delivery system choices 25–6
  dual-use options aim 29
    leave NPT or remain? 34, 35
    proliferation strategies 27–9
    routes to deterrent capability 27–8
nuclear materials
  plutonium separation 38
  polonium work 33
  technology sought 30
  see below uranium enrichment plants
nuclear weapons program
  apparent non-existence of 15, 34–5
  Chinese blueprints possibility 33
  compatible missiles sought 25
  dual-use aspirations 15
  IAEA's lack of evidence 11, 12
  materials/technology imports 12, 14
  motivation factors 11
  as national security threat 34, 36
  nuclear capability estimates 30–2, 33–4, 36, 80, 106
  opportunity costs of 36
  outside help 33
  potential "break-out" capability 35
  routes to nuclear capability 27–9
oil
  domestic usage 67
  export threats 66, 68, 82
  exports to China 134
  production 85
  as US policy driver 115–16
  referred to UN Security Council (2006) 38, 48, 124, 129, 133
religious apathy 85
Shahdom
  cooperation with Israel 93, 94
  nuclear program 67
"Shia magnetism" 72, 73, 78, 91
space program 72
United States
  "dual containment" policy 96–7
  military action possibility 80–2
  mutual animosity 11, 67–8, 69, 70, 90, 115
  regime change objective 84, 113, 114
  "tyrannical regime" view 114–15
unpopular leadership 91
uranium enrichment plants
  Arak 44
  Esfahan 44–5, 45, 61
  Iran demands "right" to 41
  Natanz 30, 31, 45, 47, 61, 131, 133
  UF4 experiments 44, 45
  uranium hexafluoride (UF6) manufacture 30, 44, 45
WMD-free Middle East proposed (1974) 137
see also EU-3; international community; sanctions
Iran Contra scheme 94
Iran, Shah of *see* Pahlavi, Mohammad Reza
Iran–Iraq War (1980)
  Iran's outmoded equipment 70
  Iran's defeat 68, 86
  Iraqi chemical weapons 86–7
  Israeli support for Iran 94
  Saddam begins 63

US support for Iraq 67–8, 86–7
Iranian Revolutionary Guard 43, 56
   Gulf/Arabian Sea war games 71
   nuclear program control 59
   support for Palestinians 122
   as US target 119, 123
Iraq
   Al Qaeda activities 78
   chemical/missile attacks on Iran 11
   effect of disintegration on GCC 87–9
   Iranian influence 72, 79, 104, 122
   Israel
      Israeli/Kurdish alliance 93–4
      nuclear threats 102, 105
      reactor attack 94, 102, 106, 111
   Kuwait invasion 110, 116
   non-signatory of CTBT 22
   nuclear activity
      Osiraq reactor 94, 102, 106, 111
      pre-Gulf War attempts 106
      reported to Security Council 134
      WMD confusion 68, 80
   ongoing war 114
   possibility of partition 79
   possible response to Iran 51
   United States
      bases in Iraq 116
      "defeat" speculation 78–9, 104
      military interventions 68–9, 103
      need for ongoing US support 89
      regime change objective 113
      united Iraq hope 63
      victory unlikely 88
   US–Iranian talks suggestion 136
   violence 122
   see also Baathist party; Hussein, Saddam
Islam: "modernists" vs. traditionalists 64
Islamic Jihad 66
Israel
   "aggressive peace" with Arabs 97
   attack shelters 120
   CTBT signatory 22

and Egypt 104, 106, 111
fear of losing nuclear advantage 109
fears over Saudi Arabia 76
influence on US politics 83–4
Iran
   Arab–Israeli consensus hopes 97
   campaign against 98–103, 111–12
   ideological conflict with 11
   Iran–Iraq War assistance 94
   non-recognition of Israel 40
   Israel's isolation attempts 97–9
   likely military targets in 119
   military strike unlikely 81
   nuclear Iran fears 98, 99, 101, 104, 106
   possible responses to Iran 106–8
   pre-revolution ties 93, 94
   rapprochement potential 109–10
   reasons for hostility 95
   "smart proliferation" fears 35
   strike capabilities 119
   US assessment of threat 82–3
   veiled nuclear threat to 98
Iraq
   nuclear threat made 102, 105
   Osiraq reactor attack 94, 102, 106, 111
nuclear capability 26–7, 101, 105
   desire to maintain 110–11
   non-member of NPT 22
   "open deterrence" possibility 103
   regional monopoly 104, 106, 107
   "second strike" plans 101, 102, 103
   seen as strategic asset 102, 104
   US "double standard" 74, 113
   US reprocessing freeze call 138
nuclear deterrence policy 110
regional position
   need to engage in WMD control 137
   strategic doctrine 104–6
   strategic situation view 104

"strategy of the periphery" 93–4, 107
Syrian diplomacy failure 103
Ivri, David 101, 102, 103

J
Japan 50, 57
   Israeli pressure on 97
   nuclear capability 15, 26
   nuclear fuel inspections 130
   US nuclear bombing of 19, 20
Johnson administration 110
Jordan
   CTBT signed/ratified 22
   Iraqi break-up scenario 88
   Iraqi population surge 79
Jordan Valley 104

K
Kashmir 59
Keyhan 43, 61
Khamenei, Ali 44, 114
Khan, A.Q.
   enrichment technology theft 14
   nuclear dissemination network 20, 27, 30, 33, 133
Khatami, Mohammed 43, 44
Khobar Towers US barracks 64
Khomeini, Ayatollah 63
   Iran–Iraq war truce 86
   Israel/Palestine stance 46
Khuzestan 63, 82
Kissinger, Henry 110
Korea (DPRK) 27
Kurds 86, 87, 93–4
Kuwait
   continuing US presence in 89
   CTBT signed/ratified 22
   Iraq's invasion of 110, 116
   Shiite majority 88

L
Lake, Tony 96–7

Larijani, Ali 44
Lebanon 88, 122
   and Al Qaeda 78
   divided Hezbollah loyalties 91
   Israeli alliance 93
   Israel's withdrawal 109
   US difficulties in 89
Lenin, Vladimir Ilyich 114
Levy, David 98
Libya
   and Khan network 33
   nuclear breaches reported 134
   nuclear technology supplied to 14, 20
lithium-6 deuteride 17
Lugar, Richard 136

M
Madrid process 96, 98–9, 103
Maronite Christians 93
MEK *see* Mujahedin-e-Khalq
metallic uranium 17–18
Middle East
   IAEA nuclear-free zone call 109
   Israeli concessions needed 137, 138–9
   membership of NPT 22
   possible responses to nuclear Iran 34, 51
   regional security forum 51
   WMD-free zone proposition 39, 40, 51–2, 137–8
   see also Gulf Cooperation Council (GCC); Gulf region
Missile Technology Control Regime (MTCR) 23
missiles
   as delivery systems 23, 24–5
   export controls on 23, 24–5
   Iran's nuclear-carrying plans 25, 48–9
   Iran's support for Hezbollah 108–9
   Israel–Iran joint venture 94

nuclear warhead suitability 24–5
  Saudi purchases of 76
Mofaz, Shaul 112
Mont Blanc 18
Mosavian, Hussein 67
Mossadegh, Mohammad 116, 121
Moussa, Amr 74
MTCR *see* Missile Technology Control Regime
Mujahedin-e-Khalq (MEK) 84
Musharraf, Pervez 66
Muslim Brotherhood: reactions to Iran 74–5

N
Nagasaki 19, 20
Natanz plant
  building of (2004) 30
  capacity estimates 31
  operations suspended 45
  operations resumed (2006) 47, 61, 131, 133
Netanyahu, Benjamin 100, 109
Netherlands, The: nuclear capability 15
Nixon administration 107, 116
Non-Proliferation of Nuclear Weapons, Treaty on the (NPT)
  aims questioned 14
  associated IAEA safeguards 13
  dual-use concerns 13
  effectiveness fears 12–13
  Iran as a non-nuclear state 42
  Iran's possible withdrawal 28–9, 34, 35
  Iran's refusal to comply (2006) 133
  members' Iran suspicions 29
  Middle East membership 22
  peaceful use clause 13–14, 41–2
  "right" to atomic energy 41–2
  Saudi Arabia signs 76
North Korea
  nuclear program reported 134
  as "tyrannical regime" 114

US regime change objective 113, 114
NSG *see* Nuclear Suppliers Group
nuclear deterrent
  as coercion opportunity 59
  mutual deterrence 100
  open deterrence 103
  types of capability 27
nuclear enrichment plants
  dual-use nature 12, 13, 15
  European multilateral plants 128
  Iran's building of 12
Nuclear Posture Review (2001/2) 119
nuclear reactors
  accident fears 73
  cooling agents 18
  Iran's lack of 22
  Israeli destruction of Iraq's 94, 102, 106, 111
  plutonium production 18
  see also individual reactors
Nuclear Suppliers Group (NSG)
  developing states oppose 13–14
  export control list 23
  nuclear technologies export ban 13, 14
nuclear weapons
  capability vs. ability 15
  designs 19–21
  hybrid 17
  key characteristics 18
  materials production 17–18
  Iranian efforts to acquire 18
  power
    implosion designs 20
    measurement method 18–19
    thermonuclear devices 20
  restrictions on use 21
  storage problems 16
  see also fission (atomic) devices; thermonuclear (hydrogen) devices

O
Obaid, Nawaf 65
oil

Iranian export threat 66, 68, 82
Iran's domestic usage plans 67
oil embargoes as sanctions 56–7
as US policy driver 115–16
Olmert, Ehud 106, 108, 110
Oman
   CTBT signed/ratified 22
   and GCC defense 77
Operation Desert Storm 68
Operation Flower 94
Osiraq reactor 94, 102, 106, 111
Oslo agreements 138

P
P-5 plus Germany 133
Pahlavi, Mohammad Reza 67
   and Israel–Iran relations 94, 107
   nuclear interest 107
   oil commitment 116
Pakistan 11
   acquires nuclear capability (1987) 26
   Chinese design modified 33
   covert nuclear development 80
   nuclear deterrent status 27, 66
   nuclear weapons program 14
   Saudi relations with 76
   threat of extremist nuclear seizure 66
   US forces in 84
   use of nuclear deterrent against India 59
   weapon initiation method 20
Palestine
   Hamas election victory 122
   Iranian stance 46
   Saddam's support of 73
   as threat to Israel 104
Palestine/Israel 78
Paris agreement 126
Peres, Shimon 96, 97, 99, 109
plutonium
   plutonium-239 (Pu-239) 17, 18, 26
   production
      Iran seeks technology 30

Iran's separation of 38
   reactor process 18
polonium 33
Putin, Vladimir 135

Q
Qatar
   CTBT signed/ratified 22
   as US base 82, 89

R
Rabin, Yitzhak 109
   and Arab–Israeli peace 97
   attitude to Iran 95, 138
   Iraq fears 138
   nuclear threat 98
radiation: effect on humans 19
radicalization 122
radiological fallout 19
Rafsanjani, Ali Akbar Hashemi 43, 44, 83
Reagan administration 94, 116
regime change
   as Israel's solution to Iran 107
   as US policy 68, 84, 113–15, 131–2
Revolutionary Guard *see* Iranian Revolutionary Guard
Romania 134
Rowhani, Hasan
   on anti-Iranian consensus breakdown 61
   uranium conversion boasts 43–5
Russia–China summit (2004) 135
Russian Federation 11, 50, 51, 55, 57, 64
   Iran
      anger at Natanz resumption 47
      builds reactor for 30
      cooperation talks (2005/6) 129–30, 131
      diplomatic pressure on 66
      fuel supplies for Bushehr 22
      launches satellite for 72
      limited enrichment suggestion 131
      limited support for 68

opposition to enrichment in 130
  rejects cooperation offer 46
  strong ties with 134–5
Sino-Russian relations 134–5
US unilateralism rejected 135

S
Salami, Hossein 71
sanctions
  choosing best form 55–7
  Clinton waives Iran's 100
  full consequences need to be known 55
  need to be universal 54–5
  as Security Council option 133–4
Saudi Arabia
  Al Qaeda threat 78, 80
  Chinese missile purchases 76
  defense capability 77
  Iraqi break-up scenario 88
  simulated Israeli air attacks on 76
  possible nuclear developer 75–7
  relations with Pakistan 76
  security threats 64, 65
  threat of nuclear Iran 75–6
  US bases in 116
  vulnerable Shiite region 73
Saudi National Security Assessment Project 65
Schröder, Gerhard 127–8
security council *see* UN Security Council (UNSC)
separative work units (SWUs) 130
September 11 events 64, 113, 115
Shamir, Yitzhak 110–11
Shamkhani, Ali 44, 70
Shapira, Shimon 109
Sharon, Ariel
  Arab–Israeli entente view 103
  on mutual deterrence 100
  pro-Shah policy 94
"Shia crescent" 91
Shiite Muslims 63, 75

GCC minorities 73, 78
Iran's influence over 72, 73, 78, 91, 122
Iraqi expulsion of 79
'represented' by Iran 65
and Saudi Eastern Province 88
Six-Day War (June 1967) 110
"smart proliferation" 22, 28, 35–6
Sneh, Ephriam 96, 99, 100
Sokolski, Henry 32
South Africa
  enrichment difficulties 33
  nuclear arsenal 16
  weapon initiation method 20
Strait of Hormuz
  importance to US 115–16
  Iranian attacks possible 122
  possible US control of traffic 119
Sunni Muslims 65, 76
  insurgency in Iraq 78
  stance on Israel 46
  Syrian majority 88
Switzerland 97
Syria 88, 104, 122
  Golan Heights assault 105
  as Hezbollah base 91
  as Iranian ally 68, 99
  Iranian relationship with 97
  non-signatory of CTBT 22
  as terrorist base 66
  as "tyrannical regime" 114
  US regime change objective 113, 114

T
Taiwan 135
Takeyh, Ray 86–7
terrorism
  Iranian threats to West 68
  regional targets 64, 65, 78
  sectarian roots 64
  see also individual organizations
testing

[175]

as credibility requirement 28
as hybrid/thermonuclear route 28
non-nuclear explosions 33
thermonuclear (hydrogen) devices
    definition of 16–17
    deployment restrictions 20–1
    testing requirement 28
tritium 17
Tudeh party 116
Turkey 87
    Israeli alliance 93

U
UN Security Council (UNSC)
    EU-3 refrain from referral 39
    Iran referred to (2005) 38, 48, 124, 129, 133
    lack of action over Iran 45
    possible fuel offer to Iran 53
    sanctions
        possible sanctions on Iran 133–4, 136
        sanctions authorization 55
        types of sanctions 23
United Arab Emirates (UAE) 89
    CTBT signed/ratified 22
    defense capability 77
    as US base 82
United Kingdom
    as EU-3 member 39
    and Iranian enrichment 131
United States 50, 57
    bombing of Japan 19, 20
    "Bush doctrine" 113
    complicated GCC role 64
    empire-building 113–14
    global non-proliferation policy 14–15
    Iran
        "dual containment" policy 96–7
        Iran-specific nuclear policy 14–15
        lack of inducements 45, 52–3
        level of military action needed 81, 82

likely military targets 119
limited enrichment rejection (2006) 131
military action possibility 80–2, 115, 117–18, 122–3
mutual animosity 11, 67–8, 69, 70, 90, 115
need to offer regime security 53–4
nuclear consequences predicted 82–3
nuclear weapons fears 40
oil supply concerns 115–16
pressure to change course 66
reasons for militant stance 115–17
regime change priority 68, 131–2
"smart proliferation" fears 35, 37
strike capabilities 119
uranium enrichment "red line" 15
US actions hasten nuclear quest 84
US as main security threat 126
Iraq
    "dual containment" policy 96–7
    military invasions 68–9, 103
    ongoing US forces need 89
    US "defeat" speculation 78–9, 104
    victory unlikely 88–9
    war as inhibiting factor 114
Israel
    Israeli interests favoured 108, 117, 139
    need to control Israeli WMD 138–9
    nuclear "double standard" 74, 113
oil/Gulf policy 69, 115–16
possible Middle East deployment 51
regime change policy 113–15
Saudi missile fears 76
"tyrannical regimes" named 114–15
uranium enrichment

[176]

centrifuge technology 12, 13, 17
    Iran seeks to acquire 30
    Iran's acquisition of 12
    P-1 technology 133
    P-2 technology 133
    production difficulties 32–3
    theft from Urenco 14
dual-use nature 13
high-enriched uranium (HEU) 31, 32, 33, 34, 35
laser technology 30
metallic uranium 17–18
UF4 experiments 44, 45
uranium hexafluoride (UF6) 17, 30, 44, 45, 133
uranium-235 (U-235) 17, 20, 26, 35
for weapons use 17–18, 30
see also plutonium
uranium, mining of 17
Urenco
    multilateral fuel plant 128
    technology theft from 14

US Defense Intelligence Agency 30, 32
US National Intelligence Council 30
US National Security Strategy 113, 114

V

Vienna 132, 134
Vienna Convention on the Law of Treaties 22–3
Vilnai, Matan 99–100

W

Wall Street Journal 82–3
Washington Institute for Near East Policy 96–7
World Trade Organization (WTO) 126

Y

Yemen 22

Z

Zangger list 23
Zimbabwe 114